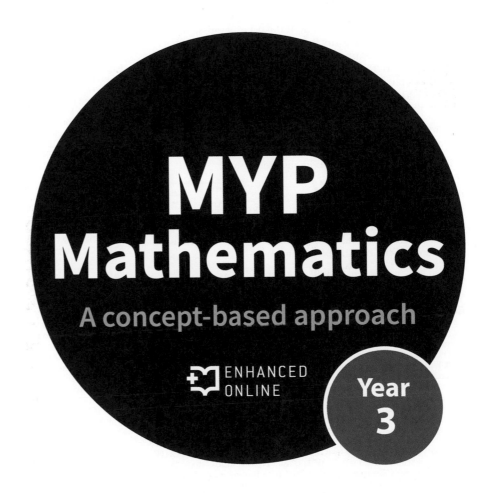

MYP Mathematics

A concept-based approach

ENHANCED ONLINE

Year 3

David Weber

Talei Kunkel

Rose Harrison

Fatima Remtulla

OXFORD

OXFORD
UNIVERSITY PRESS

Great Clarendon Street, Oxford, OX2 6DP, United Kingdom

Oxford University Press is a department of the University of Oxford.
It furthers the University's objective of excellence in research, scholarship, and education by publishing worldwide. Oxford is a registered trade mark of Oxford University Press in the UK and in certain other countries

© Oxford University Press 2018

The moral rights of the authors have been asserted

First published in 2018

British Library Cataloguing in Publication Data
Data available

ISBN 978-0-19-835617-2

10 9

The manufacturing process conforms to the environmental regulations of the country of origin.

Printed in Great Britain by Bell and Bain Ltd, Glasgow

MIX
Paper from responsible sources
FSC® C007785

Acknowledgements

Cover: Gil.K/Shutterstock

p2 (CL): 24Novembers/Shutterstock; **p2 (CR)**: Ericlefrancais/Shutterstock; **p2 (B)**: Fotokostic/Shutterstock; **p3 (TR)**: Hulton Archive/Getty Images; **p3 (Bkgd)**: Vlasov Volodymyr/Shutterstock; **p5**: Martin Freund/Shutterstock; **p6 (TL)**: Alexandr79/Shutterstock; **p6 (TR)**: Armin Rose/Shutterstock; **p6 (C)**: D. Kucharski K. Kucharska/Shutterstock; **p7**: Wikimedia; **p12**: Robert Davies/Shutterstock; **p17**: Brad Sauter/Shutterstock; **p26 (T)**: David Steele/Shutterstock; **p26 (B)**: WhiteJack/Shutterstock; **p29 (T)**: Svtdesign/Shutterstock; **p29 (B)**: Lynea/Shutterstock; **p33**: Aphelleon/Shutterstock; **p35**: 3Dsculptor/Shutterstock; **p36**: MrJafari/Shutterstock; **p37**: Raevas/Shutterstock; **p41**: Destinacigdem/123rf; **p42**: Suravid/Shutterstock; **p43**: MedusArt/Shutterstock; **p47**: Shawn Hempel/Shutterstock; **p48 (TL)**: Everett - Art/Shutterstock; **p48 (BL)**: Marco Rubino/Shutterstock; **p48 (BR)**: Teerawatyai/Shutterstock; **p49 (CR)**: Johann Knox/Shutterstock; **p49 (TR)**: WorldStockStudio/Shutterstock; **p51**: Daniel Prudek/Shutterstock; **p52 (T)**: Nikolay Antropov/Shutterstock; **p52 (B)**: Vixit/Shutterstock; **p56**: Everett Historical/Shutterstock; **p62**: Roman023_photography/Shutterstock; **p63**: Zoart Studio/Shutterstock; **p67**: Yicai/Shutterstock; **p68**: Cobalt88/Shutterstock; **p78**: Wolfgang Staib/Shutterstock; **p80**: Jag_cz/Shutterstock; **p82**: Margo Harrison/Shutterstock; **p84 (T)**: MrLeefoto/Shutterstock; **p84 (B)**: Mark_vyz/Shutterstock; **p87 (BL)**: Petch one/Shutterstock; **p87 (BM)**: Route66/Shutterstock; **p87 (BR)**: Gil C/Shutterstock; **p93**: Andrea Danti/Shutterstock; **p95**: Oleksiy Mark/Shutterstock; **p97**: Umar Shariff/Shutterstock; **p99**: Majeczka/Shutterstock; **p100 (TL)**: Shahjehan/Shutterstock; **p101 (TL)**: Ondrej Prosicky/Shutterstock; **p101 (BR)**: Songquan Deng/Shutterstock; **p103**: Fotos593/Shutterstock; **p104**: Smereka/Shutterstock; **p106**: David W Hughes/Shutterstock; **p111**: Colorcocktail/Shutterstock; **p115**: Zstock/Shutterstock; **p118**: Orlando_Stocker/Shutterstock; **p129**: Alexey Rotanov/Shutterstock; **p144**: PhotoSky/Shutterstock; **p147**: Charoenkrung.Studio99/Shutterstock; **p149**: Creative Nature Media/Shutterstock; **p151**: Frontpage/Shutterstock; **p152 (L)**: WitR/Shutterstock; **p152 (R)**: Alexander Cherednichenko/

Shutterstock; **p155**: Pecold/Shutterstock; **p156**: Daniela Mirner Eberl/NASA; **p157 (L)**: Susan Colby/Shutterstock; **p157 (R)**: National Geographic Image Collection / Alamy Stock Photo; **p158**: Steve Faber/Shutterstock; **p160**: Ricochet64/Shutterstock; **p161**: Haveseen/Shutterstock; **p162**: Globe Guide Media Inc/Shutterstock; **p164**: DeeAwesome/Shutterstock; **p167**: Jiawangkun/Shutterstock; **p172 (T)**: Picture by Watercone.com; **p172 (B)**: Akturer/Shutterstock; **p173 (TL)**: Nukul Chanada/Shutterstock; **p173 (BR)**: Fmua/Shutterstock; **p178 (TR)**: Songquan Deng/Shutterstock; **p179**: Zoran Karapancev/Shutterstock; **p181**: Timquo/Shutterstock; **p182**: dpa picture alliance / Alamy Stock Photo; **p185 (CR)**: Tatiana Liubimova/Shutterstock; **p185 (BR)**: Kolonko/Shutterstock; **p186**: Koya979/Shutterstock; **p189**: Jaroslav Moravcik/Shutterstock; **p191 (TR)**: Emiel de Lange/Shutterstock; **p194 (TL)**: SeaRick1/Shutterstock; **p194 (CR)**: Sony Ho/Shutterstock; **p194 (BR)**: Hugh Threlfall / Alamy Stock Photo; **p195 (TR)**: Dmitry Elagin/Shutterstock; **p195 (CR)**: Keith Homan/Shutterstock; **p195 (C)**: Travelview/Shutterstock; **p196 (TR)**: Climber 1959/Shutterstock; **p196 (CR)**: Daniela Mirner Eberl/NASA; **p197 (TC)**: Audrius Merfeldas/Shutterstock; **p197 (TR)**: Aleksandr Stepanov/Shutterstock; **p197 (C)**: Q DRUM SA (Pty) Ltd www.qdrum.co.za; **p197 (B)**: Vitalii_Mamchuk/Shutterstock; **p198**: Nokwalai/Shutterstock; **p202 (CR)**: Everett Historical/Shutterstock; **p202 (BR)**: Ricardo A. Alves/Shutterstock; **p203 (T)**: Everett Historical/Shutterstock; **p203 (B)**: Dominique Landau/Shutterstock; **p205**: Maxisport/Shutterstock; **p206 (TL)**: ESB Professional/Shutterstock; **p206 (CR)**: xavier gallego morell/Shutterstock; **p213**: Betto rodrigues/Shutterstock; **p214**: Think4photop/Shutterstock; **p215**: Anton_Ivanov/Shutterstock; **p216**: Rawpixel.com/Shutterstock; **p217**: Eugenio Marongiu/Shutterstock; **p219**: Monkey Business Images/Shutterstock; **p225**: Monkey Business Images/Shutterstock; **p226**: Elenarts/Shutterstock; **p230**: Monkey Business Images/Shutterstock; **p231**: The History Collection / Alamy Stock Photo; **p232**: Naluwan/Shutterstock; **p234**: Kaliva/Shutterstock; **p236**: KPG_Payless/Shutterstock; **p238**: Nice_pictures/Shutterstock; **p248**: Africa Studio/Shutterstock; **p250 (B)**: Sean Pavone/Shutterstock; **p250 (T)**: Donsimon/Shutterstock; **p252**: ARENA Creative/Shutterstock; **p254 (CR)**: Benjamin Albiach Galan/Shutterstock; **p254 (BL)**: Paul Gibbs/OUP; **p255 (T)**: Bobnevv/Shutterstock; **p255 (B)**: Nathan Jarvis/OUP; **p257**: AnnaTamila/Shutterstock; **p258 (TL)**: FooTToo/Shutterstock; **p258 (TM)**: Daniel_Kay/Shutterstock; **p258 (TR)**: Africa Studio/Shutterstock; **p258 (BL)**: S.Borisov/Shutterstock; **p258 (BR)**: Somchai Som/Shutterstock; **p259 (CL)**: Funkyplayer/Shutterstock; **p259 (CCL)**: Hunthomas/Shutterstock; **p259 (CCR)**: Funkyplayer/Shutterstock; **p259 (CR)**: Dario Sabljak/Shutterstock; **p260**: Toniflap/Shutterstock; **p261**: Khaladok/Shutterstock; **p266 (T)**: Kevin Britland/Alamy Stock Photo; **p266 (B)**: Kiev.Victor/Shutterstock; **p267 (T)**: Graphixmania/Shutterstock; **p267 (C)**: MongPro/Shutterstock; **p270**: 333DIGIT/Shutterstock; **p272**: Pawarit_s/Shutterstock; **p273 (TL)**: Ron Ellis/Shutterstock; **p273 (TR)**: Ron Ellis/Shutterstock; **p274**: Lukasz Pajor/Shutterstock; **p277 (CR)**: Sorbis/Shutterstock; **p277 (BR)**: Krisztian/Shutterstock; **p281 (CL)**: Charnsitr/Shutterstock; **p281 (CC)**: Charnsitr/Shutterstock; **p281 (CR)**: Charnsitr/Shutterstock; **p281 (B)**: Amiminkz/Shutterstock; **p282**: Artem Yampolcev/Shutterstock; **p286**: Ekkaphop/Shutterstock; **p289 (TL)**: 333DIGIT/Shutterstock; **p289 (TR)**: M.C. Escher's "Symmetry Drawing E44" © 2018 The M.C. Escher Company-The Netherlands. All rights reserved. www.mcescher.com; **p294**: Roberto Lusso/Shutterstock; **p295 (T)**: Wpadington/Shutterstock; **p295 (B)**: Zita/Shutterstock; **p297**: Cholpan/Shutterstock; **p300**: Catwalker/Shutterstock; **p303**: Romeovip_md/Shutterstock; **p306 (T)**: Stuart Taylor/Shutterstock; **p306 (B)**: Debu55y/Shutterstock; **p307 (CR)**: Smotrelkin/Shutterstock; **p307 (BR)**: Chonnanit/Shutterstock; **p308**: Topic Images Inc./Getty Images; **p309**: Santhosh Varghese/Shutterstock; **p314 (CL)**: Tarasov/Shutterstock; **p314 (CR)**: Meelena/Shutterstock; **p314 (BL)**: Mark Bassett/Shutterstock; **p314 (BR)**: Andresr/Shutterstock; **p315 (T)**: Wikipedia Commons; **p315 (B)**: Everett - Art/Shutterstock; **p317**: S_E/Shutterstock; **p318**: PathDoc/Shutterstock; **p322**: Godong/Alamy Stock Photo; **p324**: Nzescapes/Shutterstock; **p326**: Margot Petrowski/Shutterstock; **p328**: Photobyphm/Shutterstock; **p332**: Photomaxx/Shutterstock; **p334**: Jonathan Edwards; **p337**: Goran Djukanovic/Shutterstock; **p340**: Raki/Shutterstock; **p343**: Omyim1637/Shutterstock; **p344**: Andreea Dragomir/Shutterstock; **p345**: Ashley Cooper/Alamy Stock Photo; **p346**: Panda3800/Shutterstock; **p348**: Narongsak/Shutterstock; **p349**: VOJTa Herout/Shutterstock; **p352**: Dboystudio/Shutterstock; **p353**: BluIz60/Shutterstock; **p355**: Monkey Business Images/Shutterstock.

Artwork by Thomson Digital.

Contents

Launch additional digital resources for this book

MYP Mathematics 3: a new type of math textbook...

This is not your average math textbook. Whereas most textbooks present information (kind of like a lecture on paper), which you then practice and apply, this text will help you develop into a mathematician. Following the MYP philosophy, you will perform investigations where you will discover and formulate mathematical rules, algorithms and procedures. In fact, you will generate the mathematical concepts yourself, before practicing and applying them. Much like an MYP classroom, this text is supposed to be an active resource, where you learn mathematics by doing mathematics. You will then reflect on your learning and discuss your thoughts with your peers, thereby allowing you to deepen your understanding. You are part of a new generation of math student, who not only understands how to do math, but what it means to be a mathematician.

Acknowledgements from the authors:

Talei Kunkel: I would like to thank my husband, James, for his support and patience and my two children, Kathryn and Matthew, as they are my inspiration and motivation for writing this series of textbooks.

David Weber: I would like to thank my father and Bobby, the two men who inspired me to never stop believing, as well as my students, who push me to always strive for more.

Rose Harrison: To my family; David, Poppy, Beatrice, Russell and Brenda for your never-ending support and to Marlene for starting me on this incredible journey.

Fatima Remtulla: Talei Kunkel – a respected and admired colleague, who has always guided my MYP path and pushes me outside my comfort zone. Thank you for including me in this project and for showing me how much I have learned. I look forward to continuing to learn from you.

How to use this book

Each unit in this book explores mathematical content through a single global context. However, a different global context could have been selected as a focus of study. What would the study of this content look like in a different context? The chapter opener gives you just a small taste of the endless possibilities that exist, encouraging you to explore and discover these options on your own.

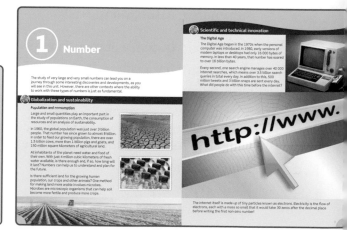

Topic opening page

Key concept for the unit.

The identified **Global context** is explored throughout the unit through the presentation of material, examples and practice problems.

Statement of Inquiry for the unit. You may wish to write your own.

The Approaches to Learning (**ATL**) skills developed throughout the unit.

You should already know how to: – these are skills that will be applied in the unit which you should already have an understanding of. You may want to practice these skills before you begin the unit.

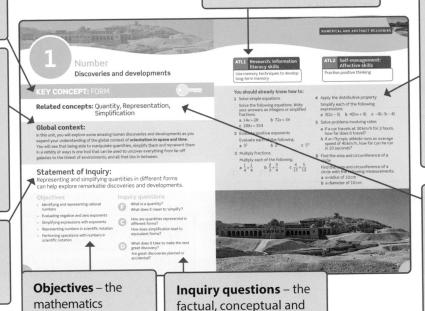

The **related concepts** explored in the unit. You may want to see if other related concepts could be added to this list.

Objectives – the mathematics covered in the unit.

Inquiry questions – the factual, conceptual and debateable questions explored in the unit.

Learning features

Investigations are inquiry-based activities for you to work on individually, in pairs or in small groups. It is here that you will discover the mathematical skills, procedures and concepts that are the focus of the unit of study.

Activities allow you to engage with mathematical content and ideas without necessarily discovering concepts. They allow you to practice or extend what you have learned, often in a very active way.

Activity 3 - Reinforcing zero and negative exponents

1 Define the term "multiplicative inverse" – research this if necessary. What happens when you multiply a number by its multiplicative inverse? Demonstrate with an example.

Hints are given to clarify instructions and ideas or to identify helpful information.

Remember: if the exponent is not explicitly written, then it is 1.

Investigation 1 – Rational numbers in different forms

criterion **B**

Take a look at the following decimal numbers:

0.5   2.7   37.4   0.3   9.12   2.4567

1 Classify each number as either "terminating/finite" or "infinite". Justify your answer.

2 Express each number as a fraction in the form $\frac{p}{q}$, where p and q are both integers and $q \neq 0$

3 Write down a general rule for how to represent a finite decimal number as a fraction.

Take a look at the following fractions:

$\frac{1}{9}$ $\frac{5}{9}$ $\frac{34}{99}$ $\frac{52}{99}$ $\frac{8}{9}$ $\frac{68}{99}$

4 Represent each fraction in decimal form. Classify each number in as many ways as possible.

5 What patterns do you notice in your answers?

6 How would you write each of the following numbers as fraction? Use a calculator to verify that each answer is correct.

a 2.4...   b $0.\overline{26}$ c 23.125125125...

7 Summarize your rules for how to represent finite decimal numbers and infinite, periodic decimal numbers as fractions.

8 Verify each of your rules for two other numbers.

9 Justify why each of your rules works.

Example 3

Q Simplify the following using the laws of exponents. Write your answer with positive exponents only.

a $9^5 \times 9^{-8}$ **b** $\dfrac{m^3}{m^{-6}}$ **c** $\dfrac{4n^2 p^{-4}}{10n^{-1}p^3}$

> The laws of exponents that you need here are the product rule and the quotient rule, both of which you have discovered in this unit.

A **a** $9^5 \times 9^{-8}$

$9^{5+(-8)} = 9^{-3}$

> When multiplying expressions with the same base, simply add the exponents.

$\dfrac{1}{9^3}$

> Rewrite your answer with positive exponents only.

b $\dfrac{m^3}{m^{-6}}$

Examples show a clear solution and explain the method.

Reflect and discuss boxes contain opportunities for small group or whole class reflection on the content being learned.

Reflect and discuss 7

- Describe two advantages of representing quantities using scientific notation.

- Which quantity is greater, 7.32×10^{12} or 4.2×10^{14}? Explain how you know.

Practice questions are written using IB command terms. You can practice the skills learned and apply them to unfamiliar problems. *Answers for these questions are included at the back of the book.*

Practice 5

1 Represent these numbers in scientific notation.

   a 23 500 **b** 365 800 **c** 210 000 000 **d** 3 650 000 **e** 569 000 **f** 7 800 000 000

2 Represent these numbers as numbers in expanded form (e.g. write 1.034×10^2 as 103.4).

   a 1.45×10^6 **b** 2.807×10^{-3} **c** 9.8×10^3 **d** 3.7×10^9 **e** 5.06×10^{-5} **f** 2×10^{-8}

3 Order the following numbers on a number line. Explain your reasoning.

   0.0025 1.42×10^4 9.83×10^{-4} 7.8×10^3 302×10^{-6} 14 2.876×10^2

4 Important discoveries in physics are listed in the table below. Represent each quantity in scientific notation.

Year	Discovery	Quantity represented as an ordinary number	Quantity represented in scientific notation
1676	Speed of light in air	299 792 km/s	km/s
1798	Acceleration due to gravity	980.665 cm/s²	cm/s²
1835	Earth's magnetic field (average)	45 microteslas	teslas
1850	Speed of light in water	225 000 000 m/s	m/s
1998	Average diameter of an atom	1 nanometer	meters

5 The word "googol" was introduced by the mathematician Edward Kasner, who asked his 9-year-old nephew what he should call the number 1 followed by 100 zeros. It is said that the company name "Google" was an accidental misspelling of the word "googol", since Google's founders planned to make incredibly large amounts of information available to people.

   a Write 1 googol in scientific notation.

Weblinks present opportunities to practice or consolidate what you are learning online or to learn more about a concept. While these are not mandatory activities, they are often a fun way to master skills and concepts.

> For practice with estimating fractions and decimals graphically, go to brainpop.com and search for 'Battleship Numberline'.

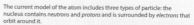

Formative assessment

The idea that all matter is composed of atoms is an incredibly important development. Before atomic theory, people had a variety of beliefs, such as the idea that all objects were made of some combination of basic elements: earth, air, fire and water. Atomic theory helps to explain the different phases of matter (solid, liquid, gas) and it allows you to predict how materials will react with each other. However, even the theory of the atom and its structure has developed over time, owing to important discoveries of the particles that make up an atom.

The current model of the atom includes three types of particle: the nucleus contains *neutrons* and *protons* and is surrounded by *electrons* that orbit around it.

Neutrons and protons have roughly equivalent masses that can be written as 16.6×10^{-28} kg. Using the same power of 10, electrons have a mass of 0.00911×10^{-28} kg.

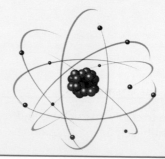

Criterion **D**

Formative assessments help you to figure out how well you are learning content. These assessments explore the global context and are a great way to prepare for the summative assessment.

Technology icon indicates where you can discover new ideas through examining a wider range of examples, or access complex ideas without having to do lots of painstaking work by hand. This icon shows where you could use Graphical Display Calculators (GDC), Dynamic Geometry Software (DGS) or Computer Algebra Systems (CAS).

ATL icons highlight opportunities to develop the ATL skills identified on the topic opening page.

ATL1

Each unit ends with

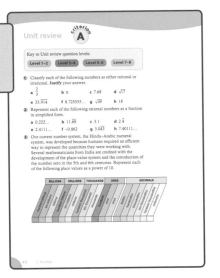

Unit summary recaps the key points, ideas and rules/formulas from the unit.

Unit review allows you to practice the skills and concepts in the unit, organized by level (criterion A) as you might find on a test or exam. You can get an idea of what achievement level you are working at based on how well you are able to answer the questions at each level. *Answers for these questions are included at the back of the book.*

The **summative assessment** task applies the mathematics learned in the unit to further explore the global context. This task is often assessed with criteria C and D.

① Number

The study of very large and very small numbers can lead you on a journey through some interesting discoveries and developments, as you will see in this unit. However, there are other contexts where the ability to work with these types of numbers is just as fundamental.

Globalization and sustainability

Population and consumption

Large and small quantities play an important part in the study of populations on Earth, the consumption of resources and an analysis of sustainability.

In 1960, the global population was just over 3 billion people. That number has since grown to almost 8 billion. In order to feed our growing population, there are over 1.5 billion cows, more than 1 billion pigs and goats, and 150 million square kilometers of agricultural land.

All inhabitants of the planet need water and food of their own. With just 4 million cubic kilometers of fresh water available, is there enough and, if so, how long will it last? Numbers can help us to understand and plan for the future.

Is there sufficient land for the growing human population, our crops and other animals? One method for making land more arable involves microbes. Microbes are microscopic organisms that can help soil become more fertile and produce more crops.

Scientific and technical innovation

The Digital Age

The Digital Age began in the 1970s when the personal computer was introduced. In 1980, early versions of modern laptops or desktops had only 16 000 bytes of memory. In less than 40 years, that number has soared to over 16 billion bytes.

Every second, one search engine manages over 40 000 internet searches, which means over 3.5 billion search queries in total every day. In addition to this, 500 million tweets and 3 billion snaps are sent every day. What did people do with this time before the internet?

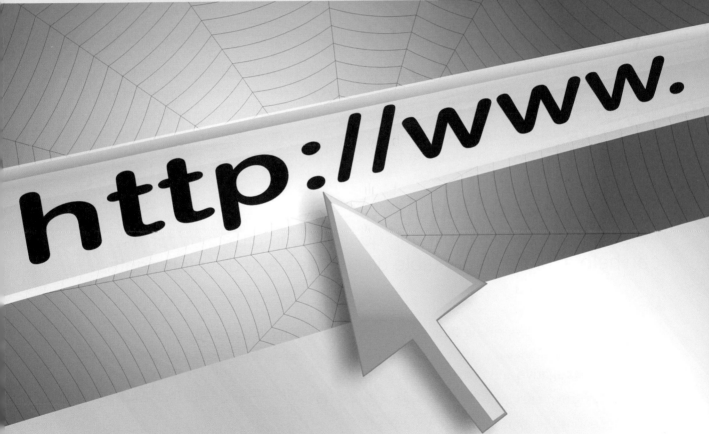

The internet itself is made up of tiny particles known as electrons. Electricity is the flow of electrons, each with a mass so small that it would take 30 zeros after the decimal place before writing the first non-zero number!

1 Number
Discoveries and developments

KEY CONCEPT: FORM

Related concepts: Quantity, Representation, Simplification

Global context:

In this unit, you will explore some amazing human discoveries and developments as you expand your understanding of the global context of **orientation in space and time**. You will see that being able to manipulate quantities, simplify them and represent them in a variety of ways is one tool that can be used to uncover everything from far-off galaxies to the tiniest of environments, and all that lies in between.

Statement of Inquiry:

Representing and simplifying quantities in different forms can help explore remarkable discoveries and developments.

Objectives

- Identifying and representing rational numbers
- Evaluating negative and zero exponents
- Simplifying expressions with exponents
- Representing numbers in scientific notation
- Performing operations with numbers in scientific notation

Inquiry questions

F
What is a quantity?
What does it mean to 'simplify'?

C
How are quantities represented in different forms?
How does simplification lead to equivalent forms?

D
What does it take to make the next great discovery?
Are great discoveries planned or accidental?

You should already know how to:

1 Solve simple equations

Solve the following equations. Write your answers as integers or simplified fractions.

a $14x = 28$ **b** $72x = 34$
c $189x = 354$

2 Evaluate positive exponents

Evaluate each of the following.

a 3^2 **b** 2^4 **c** 5^3

3 Multiply fractions

Multiply each of the following.

a $\dfrac{1}{4} \times \dfrac{1}{4}$ **b** $\dfrac{2}{3} \times \dfrac{3}{4}$ **c** $\dfrac{4}{15} \times \dfrac{5}{12}$

4 Apply the distributive property

Simplify each of the following expressions.

a $3(2x - 5)$ **b** $4(5m + 9)$ **c** $-9(-3t - 4)$

5 Solve problems involving rates

a If a car travels at 30 km/h for 2 hours, how far does it travel?

b If an Olympic athlete runs an average speed of 40 km/h, how far can he run in 10 seconds?

6 Find the area and circumference of a circle

Find the area and circumference of a circle with the following measurements.

a a radius of 10 cm

b a diameter of 10 cm

How far is really far?

How close can you get to something without actually touching it, even if it's really, really small?

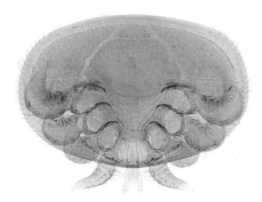

What things that exist on this planet are invisible to the human eye?

Interestingly, answering all of these questions requires the same basic skill: being able to work with numbers. From the very small to the very large, numbers are necessary to make all kinds of discoveries and developments. Whether you want to calculate how far away a new planet may be or describe the dimensions of a new strain of bacteria, you will need to know how to manipulate and perform operations with numbers represented in a variety of ways. Your ability to do this, and to describe your results accurately, may lead you to the world's next great discovery!

Rational and irrational numbers

You are familiar with integers and natural numbers, which are both types of real numbers. However, real numbers can also be classified in other ways. A number can be either *rational* or *irrational*.

Did you know...?

Our understanding of numbers has developed throughout history. We started with whole numbers since we needed to count. Fractions were developed as we discovered the need to break whole things into smaller pieces. The discovery of irrational numbers is often attributed to the Ancient Greek philosopher Hippasus, who was a Pythagorean mathematician. He used an isosceles right triangle to prove that the measure of the hypotenuse was an irrational number. It is rumoured that his discovery was so shocking that he was sentenced to death for it.

How can a number be rational or irrational?

It is said that Hippasus used a right triangle like the one shown here in his discovery of irrational numbers. In this isosceles right triangle, the sides next to the right angle measure 1 unit each, while the hypotenuse measures $\sqrt{2}$ units.

What kind of number is 1? How about $\sqrt{2}$? What makes a number rational or irrational? The activity below will help you see the difference between these two types of number.

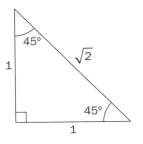

Activity 1 – What does it mean to be rational?

Some examples of rational and irrational numbers are given in the table below.

Rational numbers	Irrational numbers
0.25	0.43256931…
1.4	−1.559824…
−3.141414…	23.057419…
25.659	56.1416278…
4.5555555…	−423.745782309…
105.$\overline{37}$	72.2290004681…
−11.258258258…	2.71828182845…

1 Based on these examples, how would you define a rational number? How would you define an irrational number? Compare your definition with a peer.

▶ Continued on next page

2 Copy and complete this table. Use a calculator to represent each of the following numbers in decimal form and classify each as either rational or irrational.

Number	Decimal form	Rational or irrational?
$\frac{2}{3}$		
$\sqrt{3}$		
π		
$-\frac{3}{4}$		
$\frac{17}{99}$		
$-\sqrt{8}$		
$2\frac{2}{5}$		
$\sqrt{36}$		

> To show that a decimal repeats itself over and over again, or recurs, one of two representations can be used:
>
> 0.888888…. can also be written as $0.\overline{8}$
>
> 17.63636363… can also be written as $17.\overline{63}$

3 Based on your results in step 2, how can you recognize a rational number if it is not written in decimal form? How can you recognize an irrational number? Explain.

4 Abdisa says: "All square roots are irrational numbers." Do you agree? Justify your answer.

5 Research the meaning of the words "finite", "terminating", "infinite", "repeating", "periodic" and "non-periodic" when used to describe decimal numbers. Write down the definition of each and give an example.

6 What types of decimal numbers are rational? What types are irrational?

7 Write down two more examples (decimal, fraction, etc.) of each type of number: rational and irrational. Explain your choices.

Reflect and discuss 1

- A rational number can be defined as "any number that can be written as the ratio of two integers, as long as the denominator is not zero". How do the examples you have seen so far fit that definition?

- Is 7 a rational or irrational number? Justify your answer.

- Jeremiah says: "An irrational number is a number that cannot be written as a fraction." Is he correct? Explain.

Representing rational numbers

If a rational number can be written as a ratio of two integers where the denominator is not zero, how do you go about converting a decimal to a fraction?

Investigation 1 – Rational numbers in different forms

Criterion **B**

Take a look at the following decimal numbers:

0.5   2.7   37.4   0.3   9.12   2.4567

1 Classify each number as either "terminating/finite" or "infinite". Justify your answer.

2 Express each number as a fraction in the form $\frac{p}{q}$, where p and q are both integers and $q \neq 0$.

3 Write down a general rule for how to represent a finite decimal number as a fraction.

Take a look at the following fractions:

$\frac{1}{9}$ $\frac{5}{9}$ $\frac{34}{99}$ $\frac{52}{99}$ $\frac{8}{9}$ $\frac{68}{99}$

4 Represent each fraction in decimal form. Classify each number in as many ways as possible.

5 What patterns do you notice in your answers?

6 How would you write each of the following numbers as fraction? Use a calculator to verify that each answer is correct.

   a 2.4...   **b** 0.$\overline{26}$ **c** 23.125125125...

7 Summarize your rules for how to represent finite decimal numbers and infinite, periodic decimal numbers as fractions.

8 Verify each of your rules for two other numbers.

9 Justify why each of your rules works.

Reflect and discuss 2

- Explain why it is impossible to write a non-repeating, infinite decimal as a fraction.

- You perform a calculation on your calculator and the answer fills the screen. How do you know whether that number is a finite decimal, an infinite periodic decimal or an infinite non-periodic decimal?

- Based on the patterns you found in step 5 in Investigation 1, how should $\dfrac{9}{9}$ be represented as a decimal? How does that conflict with the value of $\dfrac{9}{9}$?

So far, you have seen how to represent finite decimals and certain kinds of periodic decimals in fraction form. What happens if the infinite decimal number has a part that repeats and a part that does not?

Example 1

Q Represent each of the following as a fraction with integer numerator and denominator.

a 0.0353535…

b 2.13789789789…

A

a
$$\text{If } x = 0.0353535\ldots$$
$$1000x = 35.353535\ldots$$
$$-10x = 0.35353535\ldots$$

Multiply the number by two different powers of 10 so that the same decimal part is repeating in both numbers.

b
$$\text{If } x = 2.13789789789\ldots$$
$$100\,000x = 213\,789.789789\ldots$$
$$-100x = 213.789789\ldots$$

$$990x = 35$$

Subtract.

$$99\,900x = 213\,576$$

$$x = \dfrac{35}{990}$$

Solve for x.

$$x = \dfrac{213576}{99\,900}$$

$$x = \dfrac{7}{198}$$

Simplify your answer if possible.

$$x = \dfrac{17798}{8325}$$

$$7 \div 198 = 0.0353535\ldots$$

Verify your answer by dividing.

$$17\,798 \div 8325 = 2.13789789\ldots$$

$$0.0353535\ldots = \dfrac{7}{198}$$

$$2.13789789\ldots = \dfrac{17798}{8325}$$

Reflect and discuss 3

- Show how you can use the procedure in Example 1 to convert 0.7777… to a fraction.

- Show how you can use the procedure in Example 1 to convert 1.242424… to a fraction.

ATL 1 & 2
- Create a two- or three-line rhyme that will help you remember how to convert decimals to fractions. Share your rhyme with a few peers and give positive feedback to each other.

ATL2
- Write down what is going well so far in this unit in terms of your learning.

Practice 1

1 Classify each of the following numbers as either rational or irrational. Justify your answer.

   a -4 **b** $\dfrac{1}{2}$ **c** 14.23 **d** $\sqrt{12}$ **e** 104.25 **f** $2.089999…$

   g $\sqrt{60}$ **h** $\sqrt{81}$ **i** $-19.34862147…$ **j** 0

2 Represent each of the following rational numbers as a fraction in simplified form.

   a 0.38 **b** $0.\overline{21}$ **c** 1.6 **d** $2.\overline{4}$ **e** 12.874

   f $0.1333…$ **g** 5 **h** -4.1392 **i** $5.6\overline{72}$ **j** $9.03555…$

3 **a** Represent $0.\overline{3}$ and $0.\overline{6}$ as fractions.

   b Show that $0.\overline{9} = 1$ by adding together $0.\overline{3}$ and $0.\overline{6}$ and then adding together their equivalent fractions which you found in step a.

   c Repeat steps a and b with $0.\overline{25}$ and $0.\overline{74}$.

   d Find your own example of two rational numbers that can be used to demonstrate that $0.999… = 1$.

Exponents

An exponent is written in the form a^x, where "a" is the base and "x" is the *exponent* or *power*. You have already seen how this can be used to simplify longer expressions. It would be difficult to write $2 \times 2 \times 2 \times 2 \times 2 \times 2 \times 2 \times 2 \times 2 \times 2$ over and over again: 2^{10} is much simpler. 2^{10} is a power of 2, with a base of 2 and an exponent of 10.

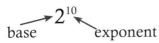

$$2^{10}$$
base   exponent

Activity 2 – The Tower of Hanoi

Mathematical puzzles have been developed over the centuries, dating as far back as Ancient Egypt. More contemporary versions include Rubik's Cube and Sudoku. The Tower of Hanoi is a puzzle invented by Édouard Lucas in 1883 that is based on a legend.

According to the legend, there is a Hindu temple that contains three large posts and 64 different gold disks. All the disks are on one post, ordered in size with the largest disk at the bottom and the smallest on top. The temple priests must move the disks one at a time from one post to another, never placing a larger disk on top of a smaller one. The puzzle is completed (and the world ends!) when all of the disks are again stacked on another post in order from largest (on the bottom) to smallest (on top). How many moves would it take to complete the puzzle?

a Simulate the puzzle using three disks to start with. You can draw the posts and disks or you can use coins to represent the disks.

 You can play an online version of the game by going to the mathisfun.com website and searching for Tower of Hanoi. The game will keep track of the number of moves and you can gradually increase the number of disks as you solve the puzzle.

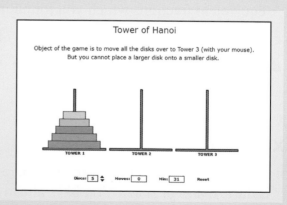

b What is the minimum number of moves you need to transfer three disks from the first post to the last, following the rule of never placing a larger disk on top of a smaller one?

▶ Continued on next page

c Gradually increase the number of disks as you solve the puzzle and record your results in a table like the one below.

Number of disks	Minimum number of moves
3	
4	
5	
6	

d Based on the pattern in your table, what is the minimum number of moves required for the priests to move all 64 of the disks to the last pole? Write your answer using an exponent.

e Use exponents to write a rule for the minimum number of moves required to solve the puzzle with n disks.

f If there were no disks, show that your rule correctly predicts the minimum number of moves necessary to solve the puzzle.

In all examples you have met before, the base has been a positive integer. What happens if the base is negative? How does that affect the value of the quantity?

Investigation 2 – Positive and negative bases

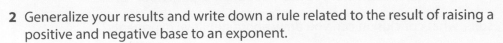

1 Find the value of each of the following.

2^2 $\qquad$ 2^3 $\qquad$ 2^4 $\qquad$ 2^5 $\qquad$ 2^6

3^2 $\qquad$ 3^3 $\qquad$ 3^4 $\qquad$ 3^5 $\qquad$ 3^6

4^2 $\qquad$ 4^3 $\qquad$ 4^4 $\qquad$ 4^5 $\qquad$ 4^6

5^2 $\qquad$ 5^3 $\qquad$ 5^4 $\qquad$ 5^5 $\qquad$ 5^6

$(-1)^2$ $\qquad$ $(-1)^3$ $\qquad$ $(-1)^4$ $\qquad$ $(-1)^5$

$(-2)^2$ $\qquad$ $(-2)^3$ $\qquad$ $(-2)^4$ $\qquad$ $(-2)^5$

$(-3)^2$ $\qquad$ $(-3)^3$ $\qquad$ $(-3)^4$ $\qquad$ $(-3)^5$

2 Generalize your results and write down a rule related to the result of raising a positive and negative base to an exponent.

3 Verify your rule for two more examples of your own choosing.

4 Justify why your rule works.

Like bases, exponents can also be integers. But what would an exponent of 0 or −1 mean?

Zero and negative powers

The typical definition of an exponent is "the number of times you multiply a base number by itself". However, exponents can be integers, decimals and even fractions. How do these unfamiliar powers fit into this very familiar definition?

Investigation 3 – Zero and negative exponents

1 What do you think 6^0 is equal to? Explain your thinking.

2 What do you think 4^{-1} is equal to? Explain your thinking.

3 Copy and complete the table by finding the value of each of the following powers. Write each answer as either a whole number or a fraction. **No decimals allowed!** If you are unsure of the value of a power, simply follow the pattern you can see in the powers above it.

$2^5 = $ _____

$2^4 = $ _____ $3^4 = $ _____

$2^3 = $ _____ $3^3 = $ _____ $5^3 = $ _____

$2^2 = $ _____ $3^2 = $ _____ $5^2 = $ _____

$2^1 = $ _____ $3^1 = $ _____ $5^1 = $ _____

$2^0 = $ _____ $3^0 = $ _____ $5^0 = $ _____

$2^{-1} = $ _____ $3^{-1} = $ _____ $5^{-1} = $ _____

$2^{-2} = $ _____ $3^{-2} = $ _____ $5^{-2} = $ _____

$2^{-3} = $ _____ $3^{-3} = $ _____ $5^{-3} = $ _____

$2^{-4} = $ _____ $3^{-4} = $ _____ $5^{-4} = $ _____

$2^{-5} = $ _____

4 Repeat the same process, starting with $\left(\dfrac{1}{2}\right)^4$ and ending with $\left(\dfrac{1}{2}\right)^{-3}$.

5 Based on your results, what conclusion can you draw about an exponent of zero?

6 Based on your results, what conclusion can you draw about negative exponents?

7 Write your conclusions as general rules, using variables instead of specific numeric examples.

8 Verify your rules for two more examples of your own, using a pattern similar to the ones above.

9 Justify why each of your rules works.

Reflect and discuss 4

- Explain how the results of Investigation 3 challenge the typical definition of an exponent.

- What do you think is the value of $\left(\dfrac{2}{3}\right)^{-2}$? Explain your answer.

 ATL1
- Create a memory aid to help you to remember your rules for zero and negative exponents.

Example 2

Q Find the value of each of the following:

a 4^{-3} **b** $(-3)^{-2}$ **c** -5^{-2}

A **a** 4^{-3}

$$\left(\dfrac{1}{4}\right)^3 \text{ or } \dfrac{1^3}{4^3}$$

$$\dfrac{1}{4}\times\dfrac{1}{4}\times\dfrac{1}{4} \text{ or } \dfrac{1\times1\times1}{4\times4\times4}$$

$$\dfrac{1}{64}$$

A negative exponent indicates that you take the reciprocal of the base.

Two fractions that multiply to 1 are called reciprocals. For example, $\dfrac{2}{3}$ and $\dfrac{3}{2}$ are reciprocals since $\dfrac{2}{3}\times\dfrac{3}{2}=1$

b $(-3)^{-2}$

$$\left(-\dfrac{1}{3}\right)^2$$

$$-\dfrac{1}{3}\times-\dfrac{1}{3}=\dfrac{1}{9}$$

A negative exponent indicates that you take the reciprocal of the base.

As you found before, a negative base raised to an even exponent produces a positive answer.

c -5^{-2}

$$-\left(\dfrac{1}{5}\right)^2$$

$$-\left(\dfrac{1}{5}\times\dfrac{1}{5}\right)=-\dfrac{1}{25}$$

A negative exponent indicates that you take the reciprocal of the base. Note that the original base is 5, not −5.

Practice 2

1 Find the value of each of the following. Write your answer as either a whole number or a fraction.

a 6^{-2} **b** $(-3)^0$ **c** 10^{-3} **d** 1^{-5} **e** $(-8)^2$

f $(-9)^{-1}$ **g** 7^3 **h** 4^{-3} **i** 5^0 **j** $(-2)^{-6}$

k $\left(-\dfrac{1}{3}\right)^3$ **l** $\left(\dfrac{3}{4}\right)^{-2}$ **m** $\left(-\dfrac{2}{11}\right)^0$ **n** $\left(\dfrac{4}{7}\right)^{-1}$ **o** $\left(\dfrac{2}{5}\right)^{-3}$

p -12^{-2} **q** $(-12)^{-2}$ **r** 0^2 **s** $-\left(\dfrac{2}{3}\right)^{-2}$ **t** $\left(\dfrac{1}{2}\right)^5$

2 Rewrite each of the following in the form a^b, where a and b are integers (no fractions).

a $4 \times 4 \times 4 \times 4 \times 4 \times 4$ **b** $\dfrac{1}{5 \times 5 \times 5 \times 5 \times 5}$ **c** $\dfrac{1}{11(11)}$

d $\dfrac{1}{10 \times 10 \times 10 \times 10}$ **e** $-7\,(-7)\,(-7)\,(-7)\,(-7)$

3 Order the quantities in each set of four from lowest to highest. Show your working.

a $5^0 \quad 4^{-2} \quad 3^1 \quad \left(\dfrac{1}{3}\right)^{-2}$ **b** $2^{-2} \quad 3^{-1} \quad \left(\dfrac{1}{4}\right)^{-1} \left(\dfrac{5}{8}\right)^0$

c $7^{-1} \quad 2^{-2} \quad 1^{-3} \quad \left(\dfrac{2}{5}\right)^2$ **d** $1^0 \quad 3^{-1} \quad 0^1 \quad \left(\dfrac{1}{2}\right)^2$

4 Ricardo says: "5^0 equals zero because you have zero 5s, which means you don't have anything. You have zero." Explain any faults in his thinking.

5 Talei says: "Negative exponents make fractions. If the negative exponent is already in the denominator of a fraction, then it makes an integer." Is Talei correct? Explain.

6 Anna says: "$8^{\frac{1}{2}}$ must be 4 since you have half of 8." Explain why this thinking is faulty.

7 Thomas states: "A positive exponent is the number of times you multiply the base. A negative exponent is the number of times you divide by the base." Do you agree? Explain your answer.

▶ Continued on next page

8 The development of the current metric system of units began in France in the 18th century. Basic units for measurements like angles, lengths, mass and capacity were created, often derived from the properties of natural objects such as water. For example, 1 liter of water has a mass of 1 kg. Multiples or divisions of these units could be created by using prefixes, such as those used in the units **milli**meter and **kilo**gram. Some of the prefixes are given in the table below. Copy and complete the table.

Prefix	Exponential form	Expanded form
giga	10^9	
mega		1 000 000
kilo	10^3	
deci	10^{-1}	
centi		$\dfrac{1}{100}$
milli		$\dfrac{1}{1000}$
micro	10^{-6}	
nano		$\dfrac{1}{1\,000\,000\,000}$
pico	10^{-12}	
femto		$\dfrac{1}{1\,000\,000\,000\,000\,000}$
atto	10^{-18}	

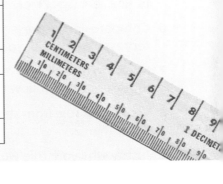

A set of 14 year old twins created the Scale of the Universe application, which allows you to see how large different powers of 10 are. Go to the site http://htwins.net/scale2/ where you can scroll to see everything from the smallest of subatomic particles to the largest of celestial objects.

Multiplying powers

You already know that 2^2 means 2×2 and 2^4 means $2 \times 2 \times 2 \times 2$. Is it possible to multiply 2^2 and 2^4 and, if so, how is this similar or different to the multiplication you do already? In the following investigations you will look for patterns and determine general rules about how to find the product of two or more powers.

Investigation 4 – Product rule

criterion B

1 Copy and complete this table.

Question	Expanded form	Simplified form
$2^2 \times 2^4$	$2 \times 2 \times 2 \times 2 \times 2 \times 2$	2^6
$3^5 \times 3^3$		
$2^4 \times 2^5$		
$y^9 \times y^2$		
$w \times w^6$		
$h^2 \times h^3 \times h^3$		

2 Describe any patterns that you see among your answers in each row.

3 Describe a rule that you could use to multiply $4^{12} \times 4^{20}$.

4 Generalize your rule to multiplying powers with any base. Be sure to use variables rather than specific numeric examples.

5 You know that 2^6 means $2 \times 2 \times 2 \times 2 \times 2 \times 2$. Rearrange this in as many different ways as possible by inserting brackets. For example, $2 \times (2 \times 2 \times 2 \times 2 \times 2)$ or $(2 \times 2) \times (2 \times 2) \times (2 \times 2)$.

6 Compare your representations with those of a peer. How many different representations can you find?

7 Knowing that all your representations should all equal 2^6, show that your representations follow the rule that you found in step 4.

8 Verify your rule from step 4 for two more examples of your own, with a base other than 2.

9 Justify why your rule works.

Reflect and discuss 5

- Can you use the rule you discovered in investigation 4 with an expression like $2^3 \times 3^4$? Explain your answer, describing any limitations on the bases involved in the multiplication.

- Can you use your rule with an expression like $4^5 \times 4^{-2}$? Explain.

- What limitations, if any, are there on the exponents? Explain.

Pairs

Activity 3 – Reinforcing zero and negative exponents

1 Define the term "multiplicative inverse" – research this if necessary. What happens when you multiply a number by its multiplicative inverse? Demonstrate with an example.

2 Write down the multiplicative inverse of 7^3.

3 Rewrite the multiplicative inverse from step 2 in the form a^b, where a and b are both integers.

4 Multiply 7^3 and its multiplicative inverse from step 3 using the product rule you have just established.

5 Repeat the same procedure for 5^6.

6 Repeat the same procedure for $\left(\frac{1}{2}\right)^4$.

7 How do your results reinforce what you have already learned about zero and negative powers? Explain.

Dividing exponents

While it is always possible to expand and multiply powers, finding a general rule allows you to accomplish the same operation much more efficiently. Is there a similar rule for dividing powers?

Investigation 5 – Quotient rule

1 Copy and complete this table.

Question	Expanded form	Simplified form	Exponential form
$2^5 \div 2^3$	$\dfrac{2\times2\times\cancel{2}\times\cancel{2}\times\cancel{2}}{\cancel{2}\times\cancel{2}\times\cancel{2}}$	2×2	2^2
$2^7 \div 2^4$			
$5^6 \div 5^2$			
$4^6 \div 4$			
$t^5 \div t^4$			
$m^8 \div m^4$			

▶ Continued on next page

2 Describe any patterns that you see among your answers in each row.

3 Generalize a rule that you could use to divide powers. Be sure to use variables rather than specific numeric examples.

4 Verify your rule for two other examples.

5 Justify why your rule works.

> Remember: if the exponent is not explicitly written, then it is 1.

Reflect and discuss 6

- Show how the quotient rule that you discovered in Investigation 5 can be used to derive the zero exponent rule.

- Show how the quotient rule can be used to derive the rule for negative exponents.

ATL2
- Write down what is going well so far in this unit in terms of your learning.

ATL2
- Identify the strengths that you have as a student. How have they helped you succeed in this unit?

Example 3

Q Simplify the following using the laws of exponents. Write your answer with positive exponents only.

a $9^5 \times 9^{-8}$ **b** $\dfrac{m^3}{m^{-6}}$ **c** $\dfrac{4n^2 p^{-4}}{10n^{-1}p^3}$

> The laws of exponents that you need here are the product rule and the quotient rule, both of which you have discovered in this unit.

A **a** $9^5 \times 9^{-8}$

$9^{5+(-8)} = 9^{-3}$

$\dfrac{1}{9^3}$

> When multiplying expressions with the same base, simply add the exponents.

> Rewrite your answer with positive exponents only.

b $\dfrac{m^3}{m^{-6}}$

$m^{3-(-6)}$

m^9

> When dividing expressions with the same base, simply subtract the exponents.

▶ Continued on next page

c $\dfrac{4n^2 p^{-4}}{10n^{-1}p^3}$

$\dfrac{2n^{2-(-1)} p^{-4-3}}{5}$

$\dfrac{2n^3 p^{-7}}{5}$

$\dfrac{2n^3}{5p^7}$

| Simplify the coefficients as you would simplify a fraction. |
| When dividing expressions with the same base, simply subtract the exponents. |
| Simplify. |
| Write your answer with positive exponents only. |

Practice 3

1 Simplify the following. Write your answers with positive exponents only.

a $5^6 \times 5^3$　　b $6^2 \times 6^{-6}$　　c $8^{-6} \times 8^2$　　d $\dfrac{x^9}{x^4}$　　e $\dfrac{y^3}{y^{-2}}$

f $\dfrac{s^8}{s^{-3}}$　　g $\dfrac{10^{-3}}{10^4}$　　h $\dfrac{12^{-4}}{12^8}$　　i $\dfrac{9^{-4}}{9^{-6}}$　　j $\dfrac{2^{-7}}{2^{-5}}$

k $\dfrac{x^{-4}}{x^6}$　　l $\dfrac{a^3}{a^{-2}}$　　m $\dfrac{w^{-11q}}{w^{-6q}}$　　n $\dfrac{m^{-7t}}{m^{-5t}}$　　o $\dfrac{y^{-3t}}{y^{8t}}$

p $\dfrac{a^{-5} \times a^3}{a^4}$　　q $5x^{-2} \times 6x^{-5}$　　r $2ab^2c^4 \times 3a^2c$　　s $(-2pq^3)(7p^3q)$　　t $(-6y^{-3}z^{-3})(-\dfrac{1}{2}y^{-1}z)$

u $\dfrac{42q^5 r^4}{6pq^4 r}$　　v $\dfrac{12d^2 ef^3}{36d^2 f^2}$　　w $\dfrac{3p^{-4}q^{-1}}{6p^2 q^{-4}}$　　x $\dfrac{45a^{-4}b^{-2}}{9ab}$

2 With the development of the metric system it became easier to compare, multiply and divide quantities because of the use of prefixes.

Prefix	Exponential form
giga	10^9
mega	10^6
kilo	10^3
deci	10^{-1}
centi	10^{-2}
milli	10^{-3}
micro	10^{-6}
nano	10^{-9}
pico	10^{-12}
femto	10^{-15}
atto	10^{-18}

a How many picograms are in a kilogram?

b How many nanobytes are in a gigabyte?

c How many femtometers are in a millimeter?

d What fraction of a megameter is an attometer?

e What fraction of a decigram is a microgram?

▶ Continued on next page

3 Which quantity is bigger? Justify each answer.

a 2^3 or 3^2 **b** 3^4 or 4^3 **c** 5^{-2} or 10^{-1} **d** 4^{-3} or 8^{-2} **e** 6^{-2} or 2^{-5} **f** 2^{-4} or 4^{-2}

4 Evaluate the following. Write your answers as integers or fractions.

a $3^{-2} \times 2^{-3} \div 6^{-2}$ **b** $7^0 \times 4^{-2} \times 2^5$ **c** $3^{-3} \div 9^{-2} \div 6^2$ **d** $\dfrac{4^2 \times 6^{-2}}{9^{-1}}$

5 In physics, dimensional analysis is often used to verify that the units in a formula produce the correct units for the answer. This process was first published by François Daviet de Foncenex in 1761, and was later developed by other scientists, for example James Clerk Maxwell.

a In 1784, Charles-Augustin de Coulomb discovered that the force of attraction between two electric charges, q_1 and q_2, can be calculated using the formula $F = \dfrac{kq_1q_2}{r^2}$, where:

> Substitute the units given into the formula and then simplify.

each electric charge is measured in coulombs (C),

r (the distance between the charges) is measured in meters,

the constant k has the units $\text{Nm}^2\,\text{C}^{-2}$.

Substitute these units into the formula and show that the formula produces the correct units of force, newtons (N).

b The electric potential (V) created by a charge q (measured in coulombs) at a distance r (in metres) from the charge was discovered to be given by the formula $V = \dfrac{1}{4\pi\varepsilon_0} \times \dfrac{Q}{r}$, where ε_0

> Remember that 4π is a constant and, therefore, has no units.

is measured in $\text{C}^2\text{N}^{-1}\text{m}^{-2}$. What should the units of electric potential be?

Formative assessment

"Those who can, teach."

criterion **C**

In this task, you will work in small groups to develop an investigation and lesson to teach your peers one of the remaining laws of exponents. The list of possible topics is given on the next page.

▶ Continued on next page

Topic	Example
Power of a product	$(xy)^m$
Power of a quotient	$\left(\dfrac{x}{y}\right)^m$
Power of a power	$(x^a)^b$
Roots as exponents	$\sqrt{x} = x^?$ $\sqrt[3]{x} = x^?$

You must hand in a detailed lesson plan which includes:

- lesson objectives/goals
- an investigation
- examples that you are going to use
- practice problems (but no word problems or problems in a global context are needed).

Objectives/goals:

- What are the desired outcomes of the lesson?
- What are the main concepts that you would like the students to understand by the end of the lesson?

Investigation:

- Create an activity that will allow your peers to discover the rule for themselves and to write it in their own words. Make sure the investigation includes at least five problems so that the pattern is clear.

Examples:

- You must guide the class through at least two examples (or as many examples as necessary to teach each type of question within your topic).
- Start with a basic one-step (single variable) question and move into multi-step questions with multiple variables.

Handout for the lesson (scaffolded notes):

- Prepare a handout to help students follow your lesson and record their learning. It should contain headings, questions and examples in the order that they will appear in your lesson. Leave blanks for students to fill in the answers to your worked examples and the key concepts that they discover during the lesson.

Practice:

- Create a worksheet to help reinforce the concepts that you have just taught. You and your group members should walk around while the class is working and help with any questions/difficulties that your peers may have.
- Provide an answer key and submit this at least one day before you teach your lesson so that your teacher can check it.

▶ Continued on next page

Activity/Game:

- In order to keep the class's attention and check their understanding, plan a game or activity for the end of the lesson. You must submit the details and questions for this activity at least one day prior to your lesson for your teacher to check.

Every group member is expected to understand the material and should be prepared to answer any and all questions from the rest of the class.

The student-created activities could be presented to the class using the ShowMe app.

Instead of a whole class activity, all of the student work can be used as a jigsaw activity where each student in a group learns one rule and teaches it to his/her peers.

Example 4

There is no order to the laws of exponents. You may use them in the order that you wish, as long as you apply them correctly.

Q Simplify the following:

a $\left(\dfrac{3i^3 j^2}{i^2 jk^4}\right)^3$

b $\left(\dfrac{4t^{-1}u^6}{18t^{-2}u^{-4}}\right)^{-2}$

A **a** $\left(\dfrac{3i^3 j^2}{i^2 jk^4}\right)^3$

Use the power of a power rule to simplify the outer exponent. Remember that the "3" has an exponent of 1.

$\dfrac{3^3 i^9 j^6}{i^6 j^3 k^{12}}$

Use the quotient rule to simplify again by subtracting exponents.

$\dfrac{27i^3 j^3}{k^{12}}$

NOTE: Alternatively, you could use the quotient rule first inside the parentheses and then use the power of a power rule.

b $\left(\dfrac{4t^{-1}u^6}{18t^{-2}u^{-4}}\right)^{-2}$

$\left(\dfrac{2tu^{10}}{9}\right)^{-2}$

Simplify using the quotient rule.

$\left(\dfrac{9}{2tu^{10}}\right)^{2}$

A negative exponent indicates that you need to work with the reciprocal.

$\dfrac{81}{4t^2 u^{20}}$

Use the power of a power rule to simplify.

Practice 4

1 Evaluate each of the following. Write your answer as either an integer or a simplified fraction.

a $\left(\dfrac{5}{9}\right)^{-2}$ b $-9^{\frac{1}{2}}$ c $(16)^{\frac{1}{2}}$ d $\left(\dfrac{3}{4}\right)^{2}$ e $\left(\dfrac{9}{10}\right)^{-1}$

f $\left(\dfrac{9}{49}\right)^{\frac{1}{2}}$ g $\left(\dfrac{1}{8}\right)^{-\frac{1}{3}}$ h $\left(\dfrac{3}{5}\right)^{-2}$ i $(-64)^{\frac{1}{3}}$ j $(81)^{-\frac{1}{2}}$

2 Simplify the following. Write your answers with positive exponents only.

a $\dfrac{8^{-2} \times 8^{-4}}{(8^{2})^{-3}}$ b $(9^{-1} \times 9^{-6})^{-3}$ c $(a^{4})^{2} \times a^{-3}$

d $\dfrac{a^{-3} \times a^{0} \times a^{4}}{(a^{2})^{-2}}$ e $\dfrac{x^{-2} y^{10} z^{-8}}{(x^{3} y^{2} z^{3})^{4}}$ f $(2t)^{3} \times (3t)^{2}$

g $\dfrac{3x^{\frac{1}{2}}}{6x^{\frac{1}{4}}}$ h $(-9y)^{2} \times (2y)^{3}$ i $-(-3x^{2})^{2}(-y^{5})$

j $\left(\dfrac{2ab^{4}}{c^{3}}\right)^{-1}$ k $2a^{\frac{1}{2}} \times 3a^{\frac{2}{3}}$ l $\dfrac{(x^{3} y^{2})(3xy^{5})^{2}}{9x^{4} y^{9}}$

m $\dfrac{(7ax^{2} z^{4})(3xy^{2})^{3}}{a^{2} zy}$ n $(4g^{3} h)^{2}$ o $\sqrt[7]{78125 a^{-7} b^{-49} c^{21}}$

p $(4a^{2} b^{3})^{-2} \times 3(a^{6} b^{3})^{\frac{1}{3}}$ q $\left(\sqrt[3]{8c^{9} d^{12} e^{15}}\right)^{2}$ r $\left(-3c^{\frac{1}{6}} d^{5}\right)^{2} \times 4(c^{2} d^{3})^{\frac{1}{3}}$

s $\dfrac{(6j^{-2} k^{3})^{-2}}{2(j^{3} k^{-1})^{3}} \times \left(\dfrac{3j}{4k^{\frac{1}{3}}}\right)^{-3}$ t $\dfrac{\left(4f^{\frac{1}{3}} g^{-1}\right)^{-3}}{2\left(\dfrac{1}{4} fg^{2}\right)^{3}}$

3 Create a bookmark with the laws of exponents on it. Make sure that all the laws fit on a small, rectangular piece of paper. Include the name of each rule and the rule itself. Enhance your bookmark through the use of colors, pictures, arrows, etc.

Scientific notation

The 17th century was a time in which very large and very small numbers were at the forefront of science. With the development of the telescope in 1610 by Galileo Galilei and with Antonie van Leeuwenhoek's improvements to the microscope roughly 50 years later, humans could now see things that were either a great distance away or were incredibly small. Being able to describe such diverse numbers with ease required the development of a new system: this is now called *scientific notation* (or *standard form*).

Writing really large and really small quantities

A water molecule measures approximately 275 picometers, where one picometer can be written as 0.000 000 000 001 m. A single drop of water contains roughly 1.7 quintillion of these water molecules. One quintillion is 1 followed by 18 zeros! Are you really supposed to write all of those zeros or is there a more efficient way to represent these quantities?

Investigation 6 – Developing scientific notation I

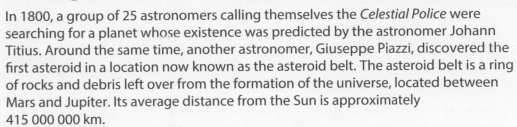

In 1800, a group of 25 astronomers calling themselves the *Celestial Police* were searching for a planet whose existence was predicted by the astronomer Johann Titius. Around the same time, another astronomer, Giuseppe Piazzi, discovered the first asteroid in a location now known as the asteroid belt. The asteroid belt is a ring of rocks and debris left over from the formation of the universe, located between Mars and Jupiter. Its average distance from the Sun is approximately 415 000 000 km.

1 415 000 000 can be represented in a variety of ways. Copy the table and fill in the missing values.

415 000 000 using products	415 000 000 using powers of 10
41 500 000 × ____	41 500 000 × 10^1
4 150 000 × ____	
_____ × 1000	
_____ × 10 000	
	4150 × 10^5
	415 × 10^6
41.5 × _____	
4.15 × _____	

2 Piazzi named that first asteroid Ceres and it is now known to be the largest asteroid in the asteroid belt. It measures approximately 950 000 meters in diameter. Use the same procedure as in step 1 to represent this quantity in a variety of ways.

950 000 using products	950 000 using powers of 10
95 000 × ____	95 000 × $10^?$
9500 × ____	
	9.5 × $10^?$

3 Look at the powers of 10 representation in the last row of each table. These are both examples of *scientific notation* or *standard form*. Describe the components of a quantity represented in scientific notation.

4 Write down a general rule for writing a large number in scientific notation.

5 The Celestial Police discovered two large asteroids in the asteroid belt: Juno and Vesta. Vesta has a mass of 259 quintillion grams. At its closest, Juno is 297 000 000 000 meters from the Sun. Verify your rule by writing each of these quantities in scientific notation.

6 Justify why your rule works.

A quantity represented in scientific notation has a decimal component, called the *coefficient*, multiplied by a power of 10.

$$2.75 \times 10^{-4}$$

coefficient   power of 10

Example 5

Q Represent the following quantities using scientific notation.

a 2139   **b** 4 980 000   **c** 310 billion

A **a** 2139

2.139×1000

2.139×10^3

> The decimal number needs to be between 1 and 10.

> Represent the power of 10 using an exponent.

b 4 980 000

4.980000×1000000

4.98×10^6

> Create a decimal number between one and ten and find the power of ten that is required to create an equivalent expression.

c 310 billion

310×10^9

$3.10 \times 10^2 \times 10^9$

3.10×10^{11}

> A third way is to represent "billion" using its equivalent as a power of 10.

> Represent 310 using scientific notation.

> Simplify using the laws of exponents.

Reflect and discuss 7

- Describe two advantages of representing quantities using scientific notation.

- Which quantity is greater, 7.32×10^{12} or 4.2×10^{14}? Explain how you know.

Did you know...?

Using the microscope he developed, Antonie Van Leeuwenhoek discovered bacteria in plaque scraped from his own teeth. He called these small organisms that measure on average 0.000005 meters "animalcules".

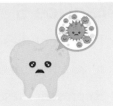

Investigation 7 – Developing scientific notation II

criterion B

1 Using the same procedure as in the previous investigation, represent 0.000006 using products of 10 and powers of 10.

0.000006 using products	0. 000006 using powers of 10
0.00006×0.1	0.00006×10^{-1}
$0.0006 \times \underline{\hspace{1cm}}$	
$6.0 \times \underline{\hspace{1cm}}$	

2 Write 0.000000304 in scientific notation.

3 Write down a general rule for writing a small number in scientific notation.

4 Van Leeuwenhoek made other discoveries as well. He was the first to see sperm cells, which have a length of 0.0047 cm, and red blood cells, whose minimum thickness is 0.8 micrometers. Verify your rule by writing each of these quantities in meters using scientific notation.

5 Justify why your rule works.

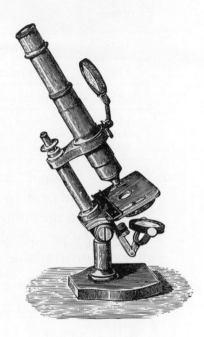

Reflect and discuss 8

- Compare and contrast scientific notation for very large and very small numbers.
- Which quantity is greater, 3.8×10^{-4} or 9.2×10^{-7}. Explain how you know.
- When a number is represented in scientific notation, how can you tell if it is less than or greater than 1? Explain.

Practice 5

1 Represent these numbers in scientific notation.

   a 23 500   **b** 365 800   **c** 210 000 000   **d** 3 650 000   **e** 569 000   **f** 7 800 000 000

2 Represent these numbers as numbers in expanded form (e.g. write 1.034×10^2 as 103.4).

   a 1.45×10^6 **b** 2.807×10^{-3} **c** 9.8×10^3 **d** 3.7×10^9 **e** 5.06×10^{-5} **f** 2×10^{-8}

3 Order the following numbers on a number line. Explain your reasoning.

   0.0025 1.42×10^4 9.83×10^{-4} 7.8×10^3 302×10^{-6} 14 2.876×10^2

4 Important discoveries in physics are listed in the table below. Represent each quantity in scientific notation.

Year	Discovery	Quantity represented as an ordinary number	Quantity represented in scientific notation
1676	Speed of light in air	299 792 km/s	km/s
1798	Acceleration due to gravity	980.665 cm/s^2	cm/s^2
1835	Earth's magnetic field (average)	45 microteslas	teslas
1850	Speed of light in water	225 000 000 m/s	m/s
1998	Average diameter of an atom	1 nanometer	meters

5 The word "googol" was introduced by the mathematician Edward Kasner, who asked his 9-year-old nephew what he should call the number 1 followed by 100 zeros. It is said that the company name "Google" was an accidental misspelling of the word "googol", since Google's founders planned to make incredibly large amounts of information available to people.

   a Write 1 googol in scientific notation.

   b Research how the quantity of information available on the internet compares with 1 googol.

   c Research "googolplex" and write it in scientific notation.

   d How do you think you would add 2 googols and 5 googols? How would you represent the operation and sum in scientific notation?

6 You have learned that a quantity written as a number in expanded form can be represented in scientific notation using a number between 1 and 10 multiplied by the appropriate power of 10. Write a six-word memory aid personal to you that will help you to remember the conversion process (e.g. "Big move left, small move right").

The development of scientific notation allows you to represent large and small numbers efficiently. How does it affect your ability to perform mathematical operations with these numbers?

Addition/subtraction with scientific notation

Our decimal system allows for numbers to be added and subtracted with ease. Do these operations become more complicated with the use of scientific notation? In this section, you will develop rules for adding and subtracting quantities represented in scientific notation.

Pairs

Activity 4 – Addition and subtraction

1 Copy this table and complete the first two columns.

Question and answer	Question in scientific notation	Using the distributive property	Answer in scientific notation
$127 + 345 = $ _____	$(\underline{} \times 10^?) + (\underline{} \times 10^?)$		$\underline{} \times 10^?$
$5212 - 3158 = $ _____			
$0.044 - 0.031 = $ _____			
$0.00025 + 0.00071 = $ _____			
$33\,208 + 19\,117 = $ _____			

2 Use the *distributive property* to rewrite the questions written in scientific notation. Write this new representation in the 3rd column.

3 Use the order of operations to evaluate the expressions in step 2. Write your answers in scientific notation in the last column.

Reflect and discuss 9

- What allows you to use the distributive property in Activity 4?

- What does this tell you about the conditions necessary to add or subtract numbers represented in scientific notation?

- What will you do if you get an answer of 14.5×10^6? How will you express this in correct scientific notation?

▶ Continued on next page

4 You are asked to add the following numbers without changing them from scientific notation.

$$(4.11 \times 10^2) + (5.08 \times 10^3)$$

What will you need to do before being able to use the distributive property?

5 Subtract the following numbers $(3.72 \times 10^5) - (2.56 \times 10^3)$ by first making sure they have the same power of 10. Express your answer in correct scientific notation.

6 Create an acronym (e.g. BEDMAS) for the process of adding or subtracting quantities represented in scientific notation. Share with a few peers, explaining what each of the letters in your acronym means.

Reflect and discuss 10

- How is adding and subtracting quantities represented in scientific notation similar to adding and subtracting fractions?

- How do you decide which power of 10 to change so that quantities written in scientific notation can be added or subtracted?

Multiplication/division with scientific notation

In 1676, the Danish astronomer Ole Rømer discovered that light travels at a specific speed that can actually be calculated. His results were based on observations of the eclipses of several moons of Jupiter by this large planet. The speed of light is approximately 3×10^8 m/s. If light travels for 1 million seconds, how far will it have traveled? Can scientific notation make multiplying these large numbers easier?

Activity 5 – How far can light travel?

Pairs

How far will light travel in 1 million seconds at a speed of 3.0×10^8 m/s?

1 Provide a reason or justification for each of the following steps.

Step	Justification
$(3 \times 10^8) \times (1 \times 10^6)$	_____
$3 \times 1 \times 10^8 \times 10^6$	_____
$(3 \times 1) \times (10^8 \times 10^6)$	_____
3×10^{14}	_____

Light will travel 3×10^{14} meters in 1 million seconds.

2 Perform the same procedure to calculate how far light will travel in 63 million seconds (approximately 2 years). Be sure to represent your answer in correct scientific notation.

3 How far can light travel in 350 000 microseconds (the blink of a human eye)? Show your working.

4 The light from the Sun takes approximately 480 seconds to reach Earth. How far does it have to travel? Use division of quantities in scientific notation to calculate this value and to represent your answer.

5 Generalize the procedure for multiplying and dividing quantities written in scientific notation.

Reflect and discuss 11

- Does scientific notation make multiplying and dividing numbers easier or more difficult? Explain.

- At what point would you represent quantities in scientific notation rather than ordinary numbers when finding their product or quotient? Explain your answer with an example.

- Write down what went well in this unit in terms of your learning.

ATL2

- What strengths in mathematics did you develop or enhance in this unit?

Practice 6

1 $a = 1.3 \times 10^6$, $b = 4.9 \times 10^{-2}$, $c = 8.32 \times 10^4$, $d = 7.6 \times 10^{-3}$, $e = 5.32 \times 10^5$

Evaluate each of the following. Represent your answers in scientific notation.

a $5a$ **b** $7b$ **c** $b + d$ **d** $2e - 4c$ **e** ae **f** bd

g $\dfrac{c}{a}$ **h** $\dfrac{e}{c}$ **i** $bc - de$ **j** $\dfrac{ac}{be}$ **k** d^2 **l** $12a - 5c + 2e$

m $\dfrac{a}{e} + \dfrac{b}{d}$ **n** $e^2 - c^2$ **o** $\dfrac{1}{d}$ **p** $2abd$ **q** $(ce)^2$

2 In the Tower of Hanoi puzzle (see pages 12–13), you found that it would take the priests $2^{64} - 1$ moves to transfer the 64 disks to a different post, maintaining their order.

  a Using a calculator, write down the number of moves using scientific notation, correct to 3 significant figures.

  b Calculate the number of seconds in one year. Represent your answer in scientific notation.

  c Legend has it that the world will end when the monks move the final disk to the new post. Suppose each move takes 1 second. Find the number of years it will take the monks to complete the puzzle if there are 64 disks. Show your working using scientific notation.

3 Considered one of the world's greatest discoveries, penicillin was actually found by accident. In 1928, Alexander Fleming returned to his laboratory to find that a sample of bacteria had been left out and had become contaminated by mold. However, where the mold had grown, the bacteria had been destroyed. A single bacterium has a diameter of 8×10^{-7} meters and a penicillium mold spore has a diameter of 3.5×10^{-6} meters.

  a How many times larger is the diameter of a penicillium mold spore than that of a bacterium?

  b Assuming a roughly circular shape, find the area of one bacterium.

  c If there are 5 million bacteria, find the area they occupy.

  d If there are 1 million bacteria and 1 million mold spores, find the total area they occupy.

  e Find the number of mold spores that would equal the area of 1 million bacteria.

▶ Continued on next page

4 The development of the rocket began over 2000 years ago with experiments and models that used steam to propel objects into the air. It was Robert Goddard who, in 1919, published a paper on how rockets could reach extreme altitudes and paved the way for the development of modern rockets. His designs and experiments formed the basis of the space programs that eventually took humans into outer space and the Moon.

a The average distance from the Earth to the Moon is 3.84×10^5 km. A rocket can travel 3.60×10^4 km on one tank of fuel. Use standard form to represent each quantity and to find the number of fuel tanks the rocket would have to take to ensure it could make the return journey.

b At its closest, Mars is 50 million km from Earth. If a rocket can travel at a speed of 5.8×10^4 km/h, how long will it take to reach Mars when at its closest distance?

c How many times further from Earth is Mars (at its closest) than the moon?

5 Satellite technology was developed in the 1960s as a means of communication, but also as a way to spy on other people. Satellites are now used for so much more, including collecting weather data, broadcasting television signals and even helping you navigate to a new destination. A satellite travels around the Earth in a circular orbit 500 kilometres above the Earth's surface. The radius of the Earth is 6375 kilometres. Calculate the maximum distance traveled by the satellite in one orbit of the Earth. Use the value of π as 3.14, or the pi button on your calculator. Write your answer in standard form with the coefficient rounded to two decimal places.

Formative assessment

criterion **D**

The idea that all matter is composed of atoms is an incredibly important development. Before atomic theory, people had a variety of beliefs, such as the idea that all objects were made of some combination of basic elements: earth, air, fire and water. Atomic theory helps to explain the different phases of matter (solid, liquid, gas) and it allows you to predict how materials will react with each other. However, even the theory of the atom and its structure has developed over time, owing to important discoveries of the particles that make up an atom.

The current model of the atom includes three types of particle: the nucleus contains *neutrons* and *protons* and is surrounded by *electrons* that orbit around it.

Neutrons and protons have roughly equivalent masses that can be written as 16.6×10^{-28} kg. Using the same power of 10, electrons have a mass of 0.00911×10^{-28} kg.

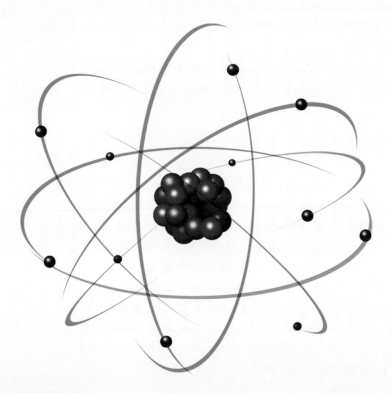

Atomic particles

a Represent each of these masses using correct scientific notation.

b Which particle has the smallest mass? Explain how you know.

▶ Continued on next page

Current elements

For each of the following questions, perform all of your calculations using scientific notation and show your working. Express your answers in correct scientific notation.

c How many times greater is the mass of a proton than the mass of an electron?

d Oxygen is an element that was discovered in 1772 by Carl Wilhem Scheele. One atom of oxygen has 8 electrons, 8 protons and 8 neutrons. Calculate the mass, in kg, of one atom of oxygen (the atomic mass). Show your working.

e The element carbon-14, which is used in dating very old objects, was discovered in 1940 by Martin Kamen and Sam Ruben. Its atomic mass is 2.324728×10^{-26} kg. If one atom of carbon-14 has 6 electrons and 6 protons, find the number of neutrons in one atom of the element. Show your working.

f Because individual atoms are so small, they are often grouped together in a larger amount, called a *mole*. The number of atoms in a mole of a substance is 6.02×10^{23}, which is referred to as Avogadro's number. If one mole of water has a mass of 18 g, find the number of atoms in 1000 grams (1 liter) of water.

> You can refer to the metric prefixes table you produced in Practice 2 question 8 (see page 17).

g The distance from Earth to Mars is 54.6 million kilometres. A helium molecule has a length of approximately 280 picometers. Find how many helium molecules could fit between these two planets.

A new discovery

h Imagine you have discovered an element unlike any other on the planet. Prepare an info-graphic that includes the following information, showing all necessary calculations using standard form:

- the name of the element and its symbol
- the number of protons, neutrons and electrons
- the properties of the element that make it so unique and valuable
- the mass of one atom of the element in grams
- the number of atoms in 25 grams of the substance.

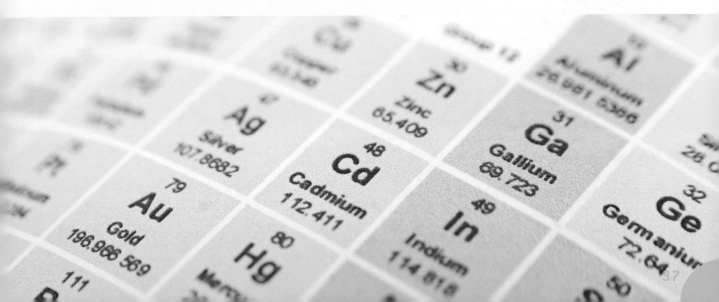

Unit summary

A *rational number* can be defined as "any number that can be written as the ratio of two integers, as long as the denominator is not zero". *Irrational numbers* cannot be written as fractions.

To convert a decimal number to a fraction:

If the decimal number is finite, read the number using its place value and write down the corresponding fraction. For example, 0.324 is "three hundred and twenty four thousandths" and can be written as $\dfrac{324}{1000}$ or $\dfrac{81}{250}$.

If the decimal number is *periodic*, multiply it by one or more powers of 10 until you have two numbers in which the decimal parts are exactly the same. Subtract them and solve the equation, as shown below:

$$\text{If } x = 0.0353535\ldots$$

$$\text{then } 1000x = 35.353535\ldots$$

$$\underline{-10x = 0.35353535\ldots}$$

$$990x = 35$$

$$x = \frac{35}{990} \text{ or } \frac{7}{198}$$

Expressions containing quantities raised to an exponent can be simplified using the following laws of exponents.

Product rules:

    Product rule with same base: $a^n \times a^m = a^{n+m}$

    Product rule with same exponent: $a^n \times b^n = (ab)^n$

Quotient rules:

    Quotient rule with same base: $\dfrac{a^n}{a^m} = a^{n-m}$

    Quotient rule with same exponent: $\dfrac{a^n}{b^n} = \left(\dfrac{a}{b}\right)^n$

Power of a power rule: $(a^n)^m = a^{nm}$

Zero power rule: $a^0 = 1$

Negative power rule: $\dfrac{1}{a^m} = a^{-m}$

Fractional exponents: $\sqrt[a]{x} = x^{\frac{1}{a}}$

Quantities represented in *scientific notation* (also known as *standard form*) have a *coefficient* between 1 and 10 multiplied by a power of 10. For example:

$$1.67 \times 10^5$$

coefficient   power of 10

To multiply or divide quantities represented in scientific notation, multiply/divide the coefficients and multiply/divide the powers of 10 using the laws of exponents. For example:

$$(2.4 \times 10^7) \times (1.5 \times 10^{-3}) = (2.4 \times 1.5) \times (10^7 \times 10^{-3})$$
$$= 3.6 \times 10^4$$

To add or subtract quantities represented in scientific notation, rewrite the quantities so that they have the same power of 10 and then add or subtract the coefficients. For example:

$$(2.4 \times 10^8) - (1.5 \times 10^7) = (24 \times 10^7) - (1.5 \times 10^7)$$
$$= 10^7 \times (24 - 1.5)$$
$$= 22.5 \times 10^7 \text{ or } 2.25 \times 10^8$$

Unit review

Launch additional digital resources for this chapter

Key to Unit review question levels:

Level 1–2   Level 3–4   Level 5–6   Level 7–8

1. Classify each of the following numbers as either rational or irrational. **Justify** your answer.

   a $\frac{2}{3}$ b π c 7.68   d $\sqrt{17}$

   e $33.9\overline{14}$ f 8.725555…   g $\sqrt{49}$ h 18

2. Represent each of the following rational numbers as a fraction in simplified form.

   a 0.222…   b $11.\overline{68}$ c 3.1   d $2.\overline{4}$

   e 2.4111…   f −0.862   g $3.0\overline{43}$ h 7.40111…

3. Our current number system, the Hindu–Arabic numeral system, was developed because humans required an efficient way to represent the quantities they were working with. Several mathematicians from India are credited with the development of the place-value system and the introduction of the number zero in the 5th and 6th centuries. Represent each of the following place values as a power of 10.

BILLIONS			MILLIONS			THOUSANDS			ONES			DECIMALS				
hundred billions	ten billions	billions	hundred millions	ten millions	millions	hundred thousands	ten thousands	thousands	hundreds	tens	ones	tenths	hundredths	thousandths	ten thousandths	hundred thousandths	millionths

4 Evaluate each of the following. Write your answer as either an integer or a simplified fraction.

a 5^{-3}

b 3^{-3}

c $(-27)^{\frac{1}{3}}$

d $\left(\dfrac{3}{4}\right)^2$

e $\left(\dfrac{2}{5}\right)^{-1}$

f $\left(\dfrac{4}{25}\right)^{-\frac{1}{2}}$

g $\left(\dfrac{1}{64}\right)^{-\frac{1}{3}}$

h $\left(\dfrac{6}{7}\right)^{-2}$

5 Simplify each of the following. Express your answers with positive exponents only.

a $a^3b^2 \times a^4b^{-5}$

b $(28x^2y^{-3})^0$

c $-(14m^2n^3)(-2m^3n^2)$

d $\dfrac{9a^2}{(-3a)^2}$

e $(6x^5y^{-2})(3x^2y^3)$

f $\dfrac{-14x^6y^7}{7x^{-3}y^9}$

g $(6x^4y^3)(-4x^{-8}y^{-2})$

h $(15x^{4c})(7x^{-6c})$

i $\dfrac{20n^4m^{-3}}{8n^8m^{-5}}$

6 Find the following products and quotients. Write your answers using scientific notation.

a $(3 \times 10^5) \times (4 \times 10^8)$

b $(2.4 \times 10^9) \div (1.2 \times 10^6)$

c $(2.5 \times 10^{-4}) \times (3.1 \times 10^{-3})$

d $\dfrac{9.2 \times 10^3}{4.2 \times 10^6}$

e $(8.1 \times 10^{-2}) \div (6.8 \times 10^{-7})$

f $(6.2 \times 10^{11}) \times (4.9 \times 10^{-13})$

7 Our understanding of our own place in the universe has developed over time, often influenced by personal beliefs rather than scientific evidence. While Nicolaus Copernicus is credited with formulating the current model of our solar system, a Greek astronomer, Aristarchus, promoted the idea of the Sun being at the center of our universe about 1800 years earlier.

Johannes Kepler was the first to propose the laws of planetary motion that are still in use today.

The distance between the Earth and Mars is constantly changing as the planets rotate about the Sun. The smallest distance between them is 54.6 million km and the greatest distance between them is 401 million km. The fastest spaceship to leave Earth was NASA's *New Horizons* with a recorded speed of 58 000 km/h. Use this information to answer the following questions.

a **Calculate** the minimum number of days the spaceship would take to travel from Earth to Mars when the planets are at their closest to each other.

b **Calculate** the minimum number of days the spaceship would take to travel from Earth to Mars when the planets are at their furthest apart.

8 Find the following sums and differences. Write your answers using standard form.

a $(6 \times 10^8) - (2 \times 10^7)$

b $(5.5 \times 10^{-2}) + (3.1 \times 10^{-4})$

c $(7.3 \times 10^{-4}) - (8.6 \times 10^{-3})$

d $(6.27 \times 10^3) + (5 \times 10^4)$

e $(9.1 \times 10^{11}) + (4.4 \times 10^{13})$

f $(5 \times 10^9) - (2.7 \times 10^7)$

9 The Human Genome Project began in 1990 with the goal of discovering and recording the complete sequence of DNA base pairs in human genetic material. It was a 13-year project that was almost completely successful, mapping over 99% of the human genome.

The table below gives the number of base pairs mapped during each three-year period from 1990 to 1999.

Years	Number of base pairs mapped in each period
1990–1993	2.0×10^6
1993–1996	1.4×10^8
1996–1999	4.7×10^9

a Find the total number of base pairs that were mapped between 1990 and 1999.

b If there are 3.2×10^{10} base pairs in the human genome, find the number of base pairs that were mapped in the final four years of the project.

10 Simplify each of the following. Express your answers with positive exponents only.

a $\left(\sqrt{49\,p^{26}q^{10}r^{-12}}\right)$

b $\dfrac{\left(2u^{\frac{2}{3}}v^{-2}\right)^{-6}}{(uv^3)^2}$

c $\dfrac{(a^5b^{-6})^{-2}}{(a^0b^7)^3}$

d $\left(\dfrac{x^2y^6}{a^{-2}b}\right)^2\left(\dfrac{x^3y^0}{ab}\right)^{-3}$

e $(16a^4b^{-2}c^6)^{\frac{1}{2}}(-27a^{-6}b^9c^3)^{-\frac{1}{3}}$

f $\left(\dfrac{36m^3n^0}{121m^{-4}n^{-5}}\right)^{\frac{1}{2}}$

g $(25g^{-5}h^{-1}j^7)^{\frac{1}{2}}(-8g^4h^{-2}j^{-1})^{\frac{1}{3}}$

11 Neuroscience is the study of the nervous system, with neuroscientists focusing on the brain and its structure. It is a discipline that has developed since 500 BC, with major advances being made since the middle of the 20th century. While the brain has a high composition of water and fat, it also contains a large number of neurons that transmit and receive signals.

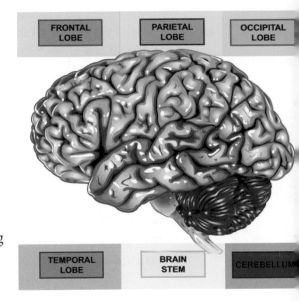

The cerebellum contains roughly 69 billion neurons. The cerebral cortex, made up of the frontal, parietal, occipital and temporal lobes, contains roughly 16 billion neurons. The remaining structures contain 690 million neurons.

a How many neurons are there in the brain in total? Express your calculations and answer in scientific notation.

b It has often been said that the number of neurons in the brain is the same as the number of stars in the universe. If there are 7.0×10^{22} stars in the universe, is this statement true? If not, which quantity is greater and by what factor?

c Each neuron has a length of approximately 1×10^{-4} meters. If they were placed end to end, what would be the length of all of the neurons in the brain?

12 The Richter Scale was devised by Charles Richter in 1940 to compare the intensities of earthquakes. The intensity of an earthquake is determined by the amount of ground motion measured on a seismometer. Each increase of 1 unit in magnitude on the Richter scale corresponds to an increase of 10 times the intensity measured on a seismometer.

a Using this ratio, how many times more intense was the 1556 earthquake in China with a magnitude of 8 compared with the 2010 earthquake in Haiti with a magnitude of 7?

By expressing the intensity (I) as an exponential function of the magnitude (M), you can compare the intensities of earthquakes that do not differ by a whole integer.

$$I = 10^M$$

b The world's most powerful earthquake was in Chile in 1960 and registered magnitude 9.5 on the Richter Scale. The deadliest recorded tsunami was caused by an earthquake which registered magnitude 9.1 on the Richter scale off the coast of Indonesia in 2004. Using the formula given above, how many times more intense was the earthquake in Chile than the one in Indonesia?

c The two most costly earthquakes both occurred in Japan: the earthquake of 2011 had a magnitude of 9.1; the earthquake of 1995 had a magnitude of 6.9. Using the formula given above, how many times more intense was the earthquake in 2011 compared with the one in 1995?

Summative assessment

Microchip technology

How is it possible to surf the internet? How can a smartphone control so many devices? How does a pacemaker help control a heart's contractions? At the core of all of these is a single device called a *transistor*. The transistor was invented in 1945 in Bell Labs and the inventors had little idea how much it would revolutionize our way of life. In this task, you will analyze the growth of transistor technology and the development of the microprocessor chip.

You will present your work for each part in a single report. Show your working in each section. Perform all your calculations and write all your answers using scientific notation.

Part 1 – Moore's law

Gordon Moore, one of the founders of Intel, helped build a company that produces processors for computer manufacturers. Processors, or microprocessors, are small chips inside devices such as smartphones and computers that receive input and produce output using transistors. In what has been named Moore's law, Moore predicted that the number of transistors that would fit on a chip would double every two years.

a   If the very first chip had four transistors, use Moore's law to calculate the number of transistors on a chip every two years over the next 10 years. Copy and complete this table, writing your answers as powers of 2.

Year	Number of transistors on chip
0	4
2	$4 \times 2 = 2^?$
4	
6	
8	
10	

b If there were four transistors on a chip in 1965, predict the number of transistors on a chip in the year 2015. Write your answer both as a number in expanded form and in scientific notation.

c During a speech in 2014, one of Intel's vice presidents said that, by 2026, the company would make a processor with as many transistors as there are neurons in a human brain. If there are 1.0×10^{11} neurons in the human brain, would Moore's law agree with the vice president's statement?

Part 2 – Chip technology

Transistors can perform two functions. They can amplify current so that an input current is greatly increased as it passes through the transistor. Because of this, transistors were originally used to develop hearing aids. They can also act as switches, being either "on" or "off". This allows the transistor to store two different numbers, either a 0 (off) or a 1 (on). Originally, vacuum tubes were used as switches, but these were large and required a lot of power.

a Smartphones have chips in them that can contain 3.3 billion transistors. If each transistor weighs 5.1×10^{-23} grams, find the total mass of the transistors in a smartphone.

b If each chip has a length of 35 nanometers (nm), how many would you need to circle the Earth, which has a radius of 6371 km?

c Supercomputers have been developed that are much larger and can perform many more calculations than ordinary desktop or laptop computers. One such supercomputer, the Titan, has 4.485×10^{10} transistors in its central processing unit (CPU) and another 1.3268×10^{11} transistors in its graphics processing unit (GPU). Find the total number of transistors in the Titan. Show your working and give your answers in standard form.

d Intel estimates that about 12 quintillion transistors are shipped around the globe each year. If that represents 10 000 times the number of ants on the planet, find the number of ants on Earth.

Part 3 – Design your own

What if you could design your own processor? How small would you make it? How many transistors could you fit on it?

The size of transistors has decreased dramatically since they were first invented. Assume you will use transistors that are approximately rectangular and measure 35 nm by 14 nm.

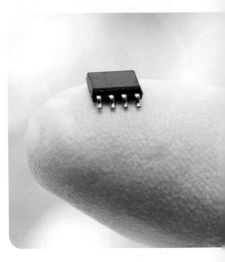

a Select a chip size that sounds impressive (e.g. a fingernail). Find its area. (You may choose to research the area or calculate it after taking measurements.)

b Find the number of transistors that you will be able to fit on your chosen area.

c If transistors costs $0.000000003 USD each, find the cost of the transistors on your chip.

d Create a headline to announce your technology to the world.

e Write a newspaper article about your invention and create a snazzy name for your chip. Your article must include the following:

- **Headline** – usually only a few words. It's purpose is to attract the interest of the reader by giving a hint as to what the article is about in a concise way.

- **By-line** – the author of the article.

- **Introduction** – sets the scene and summarizes the main points of the article: *who, what, when, where*.

- **Body** – provides more detail about the event, in particular it answers the questions *how* and *why*.

- **Quotes** – what a person (such as an eye-witness or an expert) has said about the invention. These will be in speech marks.

- **Photograph and caption** – include a drawing or photograph of your invention as well as a caption that describes what is in the photo.

- **Answers to these questions** – What does it take to make the next great discovery? Are great discoveries planned or accidental?

(2) Triangles

Many scientific principles, processes and solutions rely on the triangle, as you will discover in this unit. However, triangles have also played important roles in other contexts that could have taken your study in a completely different direction, such as our appreciation of the aesthetic and urban planning and infrastructure.

Personal and cultural expression

Appreciation of the aesthetic

While you might think that great art is 100% inspiration and creativity, some of it involves very careful planning in order to produce the desired effect. A study of triangles could be a lesson on how mathematical principles can enhance our appreciation of the aesthetic.

During the Renaissance, artists realized the power of the triangle in composing a work. Leonardo da Vinci used the Golden Triangle to bring attention to the Mona Lisa's face in his famous painting. The Golden Triangle was also supposed to produce an aesthetically pleasing portrait.

Photographers use triangles to give a sense of stability to their work. Can you see the triangle in this picture? How does it make the scene appear stable? Is there a particular type of triangle that produces the optimum effect?

Artists also use triangles to give a sense of perspective to a work. How does a triangle accomplish that in this image?

Urban planning and infrastructure

Cities are often planned using a rectangle (grid) or a circle (radial). However, with an ever-increasing population, urban planners are becoming more creative with their use of limited space. A study of triangles could take you on a tour of some very innovative urban structures and designs.

This housing development in northern Spain utilizes a triangular shape to provide living spaces that have both interior and exterior views as well as a central gathering space.

In the Philippines, the Ayala Triangle Gardens use the space between three major roads to offer citizens an alternative to walking alongside the busy streets. Its streets and paths also offer a shorter route than the bordering roads.

2 Triangles
Principles, processes and solutions

Related concepts: Generalization, Measurement

Global context:

How was the engine invented? How does solar power work? These are examples of how a basic scientific principle (the law of conservation of energy) was used to create new products, processes and solutions. In this unit, your exploration of the global context of **scientific and technical innovation** will lead you on a journey of how scientists and mathematicians use the foundational principles in their disciplines to formulate new methods and ideas. These new ideas can then be applied to solve problems that threaten the environment or plague our everyday lives.

Statement of Inquiry

Generalizing relationships between measurements can help develop principles, processes and solutions.

Objectives

- Solving problems involving right triangles using Pythagoras' theorem
- Determining whether or not two triangles are similar
- Using the properties of similar triangles to find missing measurements
- Using trigonometric ratios to solve problems involving right triangles

Inquiry questions

F
What is a relationship?
What does it mean to measure something?

C
How do we generalize relationships between measurements?

D
How much proof is "enough"?
Do scientific principles lead to good solutions or do good solutions lead to scientific principles?

ATL1 Thinking:
Critical-thinking skills

Test generalizations and conclusions

ATL2 Communication:
Communication skills

Give and receive meaningful feedback

You should already know how to:

1 Find the areas of common shapes

Find the area of each shape.

a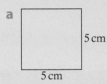

b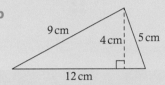

c

2 Solve multi-step equations

Solve the following equations.

a $3x^2 = 48$ b $\dfrac{x}{7} + 2 = 7$

c $\dfrac{5+x}{9} = 1$ d $5x^2 + 6 = 51$

3 Plot points on a coordinate grid

Plot the following points.

a $(-3, 4)$ b $(0, -2)$
c $(-1, -6)$ d $(8, 2)$

4 Use properties of parallel lines to determine sizes of angles

Find the size of the indicated angles. Justify each step.

a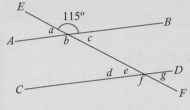

b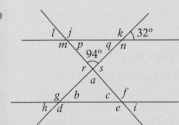

5 Solve problems using ratios

a $\dfrac{12}{28} = \dfrac{x}{21}$ b $8:11 = 17:m$

c $\dfrac{10}{x} = \dfrac{17}{15}$

51

Introducing triangles

What do kitchens, satellites and Mount Everest all have in common? Interestingly enough, triangles!

The *kitchen work triangle* is a concept used when designing kitchens. Since the sink, refrigerator and stove are the main centres of activity, the optimal design of a kitchen is based on a triangle between these appliances. There are even optimal measurements to minimize the walking distance between these areas.

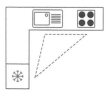

L-shape kitchen

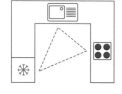

U-shape kitchen

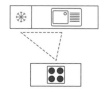

Island kitchen

While orbiting above Earth, satellites have a clear view of a limited portion of the planet. This range can be calculated using right triangles. Cameras, such as those used in taking aerial photos, apply triangle principles to produce photos with the right amount of detail.

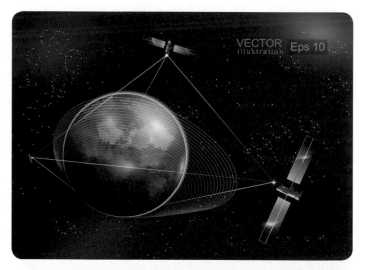

Even solving the problem of how to determine the height of locations that are very difficult to access, like Mount Everest, can be easily accomplished with the right tools and a knowledge of triangles.

In this unit, you will discover how an understanding of the relationships between measures in triangles has allowed humans to develop some pretty clever principles, processes and solutions. Who knows, you may make the next great discovery as you investigate this common shape.

Reflect and discuss 1

- How easy is it to move between the stove, sink and refrigerator in your home? Does your kitchen have a kitchen work triangle? If so, what kind of triangle is it?

- What processes or solutions have you seen that use triangles? Give two examples.

- Do you think scientific principles lead to good solutions or do good solutions lead to scientific principles? Explain.

Theorems and proof

In mathematics, an *axiom* or *postulate* is a statement that is assumed to be true and, therefore, needs no proof. For example, the mathematician Euclid stated several axioms, such as: "Things which are equal to the same thing are also equal to one another". This may seem a rather obvious statement, but it forms the basis of many other mathematical principles. A *theorem*, on the other hand, is a statement that has been *proved*, often with the help of established axioms. Theorems start out as *conjectures* (educated guesses) until they are proved. But what does it take to prove something? How much evidence is enough?

Proof

Mathematics is based on deductive reasoning, which means coming up with new statements based on true statements. Proofs are how mathematicians construct knowledge and know that it is correct.

In mathematics, a *proof* is a logical sequence of arguments whose goal is to convince people that something is true. Unlike in a court of law, "reasonable doubt" is not acceptable. Every step of the proof must be proved to be true for every case, without **any** doubt.

Activity 1 – Is a pattern enough proof?

1 Looking at the diagrams above, copy and complete this table.

Number of points on the circle	Number of different regions in the circle
1	
2	
3	
4	
5	

2 What do you predict will be the number of regions if there are six points on the circle? What about seven points? Explain your reasoning in each case.

3 Write down a conjecture predicting how many regions there will be in any circle as compared with the number in the previous circle.

4 Do your results in this activity **prove** the statement you wrote in step 3? Explain.

ATL1

5 Look at the diagrams below and count the number of regions in the circle when there are six points and when there are seven points. Add two rows to your table for these results.

6 Was your conjecture true? Explain.

7 Is establishing a pattern enough proof? Explain.

> A conjecture is a conclusion or educated guess that is based on the information that you have.

> Try drawing larger copies of these two diagrams, so you can number the regions as you count them.

Activity 1 is a good example of why you cannot say something is "always true" in mathematics based on just a few examples. Proving something to be true has to start from basic principles that are known and accepted facts.

What is a theorem?

A *theorem* is a mathematical statement that can be proved to be true. The mathematics that you study is based on centuries of work, but new theorems are still being developed. In 2016, a new theorem related to circles was established by an Israeli teenager as she did her math homework! She later proved her theorem, which now bears the young mathematician's name: "Tamar's theorem". As you go through this unit, you will encounter examples of both theorems and proofs.

Activity 2 – Four-square theorem

In 1770, Joseph Louis Lagrange published the proof of a theorem that would eventually be called the *Lagrange four-square theorem*. It states, "Every positive integer can be written as the sum of the squares of four integers."

For example, $35 = 1^2 + 3^2 + 3^2 + 4^2$ or $11 = 3^2 + 1^2 + 1^2 + 0^2$

1 Write the following numbers as the sum of the squares of four integers.

   a 17   **b** 30   **c** 43   **d** 103

2 Select any positive integer and write it as the sum of the squares of four integers.

Pairs

3 With a peer, play the following game. Each person selects a 2-digit number that the other player must write as the sum of the squares of four integers. Score one point for every time you solve the problem and one point for every time you create a problem that cannot be solved. Add a time restriction to make the game more challenging.

Reflect and discuss 2

- Have you **proved** the four-square theorem with your examples? Justify your answer.

- The formula for the area of a circle is actually a theorem proved by the Greek mathematician Archimedes. It states, "The area of any circle is equal to the area of a right-angled triangle in which one of the sides about the right angle is equal to the radius, and the other to the circumference of the circle." How does this relate to what you already know about the area of a circle?

- How difficult do you think it is to prove mathematically a statement like this one by Archimedes? Explain.

One of the most famous theorems that you will learn is Pythagoras' theorem. It establishes a relationship between the measurements of the sides in any right triangle.

In a right triangle, the sides have specific names. The side across from the right angle, the longest side, is called the *hypotenuse*. The other two sides, which form the right angle, are called *legs*.

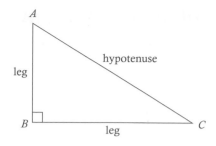

Reflect and discuss 3

- Explain why the side across from the right angle has to be the longest side.

- Would you say that you have "proven" this result? Explain.

- Where is the shortest side in a triangle located? Be as specific as you can.

Pythagoras' theorem

In the activities in this section, you will explore the relationship found by Pythagoras and examine some of the proofs associated with his theorem.

Did you know...?

Pythagoras was a Greek mathematician born on the Island of Samos in around 569 BC.

His most famous theorem explained the relationship between the lengths of the sides in right-angled triangles. Although he never wrote down any of his findings, his followers and students did later on.

Pythagoras' theorem is possibly the most proven of all principles. It can be proved both visually and algebraically.

Activity 3 – Pythagorean puzzles

Your teacher will give you the Pythagorean puzzle cutouts.

You can also download and print this puzzle from the following website:
www.jamieyorkpress.com – Select "Free Downloads" and scroll down the "Grade 7 Downloads", where you will find the "Pythagorean Theorem Cutout Puzzle".

1 Cut out the pieces from each of the smaller squares and rearrange them to form a square on the hypotenuse.

2 What generalization can you deduce about the relationship between the area of the squares on each of the sides of a right triangle?

ATL2

Pairs

3 With a partner, look at each other's generalizations. Take turns giving feedback by first asking questions to clarify anything that isn't quite clear. Then, give at least one positive comment. Follow that with a suggestion for improvement, if appropriate.

The previous activity is a good first step in generalizing the relationship between the sides of a right triangle. But what exactly is Pythagoras' theorem? Breaking up the squares into much smaller ones, as you will see in the next activity, will help you to discover this famous relationship.

Investigation 1 – Discovering Pythagoras' theorem

criterion B

You can use paper and pencil or dynamic geometry software to perform the following activity.

You can use the applet on the learnalberta.ca website instead of drawing the triangles needed for this investigation. In the "Find Resources" section of the home page, enter the search term "Exploring the Pythagorean Theorem" in the "Enter Keyword" box. Then click on the link that appears on the right of the page. Select the "Interactive" option and choose the size of triangle you want.

1 In the middle of a piece of squared paper, draw a right triangle whose legs measure 6 units and 8 units.

▶ Continued on next page

2 Construct squares on these two legs. (The length and width of the square is the length of the side of the triangle.)

3 Find the number of small squares in these two new squares. What does this represent?

4 Construct a new square on the hypotenuse.

5 Count the total number of squares in this larger square. You may need to count "half squares" together.

6 What do you notice about the area of this larger square? How does it relate to the area of the two smaller squares?

7 Repeat steps 1 through 6 for a right triangle with legs that measure 5 and 12 units.

8 Repeat steps 1 through 6 for a right triangle with legs that measure 8 and 15 units.

9 If the measures of the legs were a and b, how would you calculate the area of the square formed on each leg of the squares drawn on these sides?

10 Write down Pythagoras' theorem as a relationship between the areas of the three squares.

 11 Verify that your formula works for two more right triangles:

  a a right triangle with legs of length 7 and 24 units

  b a right triangle with legs of length 20 and 21 units.

12 Justify why your formula works.

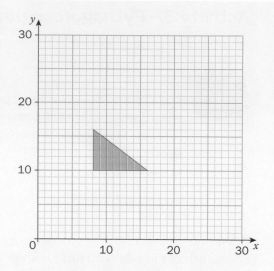

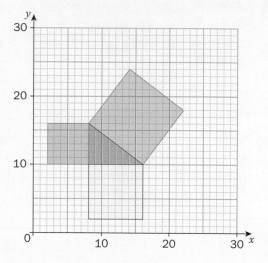

Reflect and discuss 4

Answer these questions on your own and then discuss them in a group.

- Explain how Pythagoras' theorem generalizes a relationship between measurements.

- Show that Pythagoras' theorem does not work for triangles that are not right-angled. Explain your method.

- Most people remember Pythagoras' theorem as $a^2 + b^2 = c^2$. What do the a^2, b^2 and c^2 represent? How do you know which side of the triangle is isolated in the theorem?

> A term that is isolated is by itself on one side of the equation.

There are over 350 proofs of Pythagoras' theorem! Some involve algebra while others are visual proofs.

Activity 4 – A visual proof of Pythagoras' theorem

A visual proof of Pythagoras' theorem using two squares is shown below.

For an animation showing this visual proof, go to youtube and search for "Pythagoras' Theorem Proof Animation" and select the video by Maths Whenever.

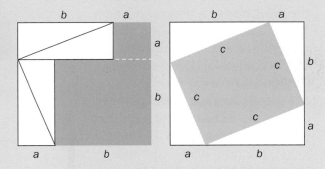

1 Explain how you know that both of the large squares have the same area.

2 Write down an expression to calculate the area of the green square in the diagram on the right.

3 Write down an expression to calculate the area of each of the two blue squares in the diagram on the left.

4 How do you know that the sum of the areas of the blue squares equals the area of the green square? Explain.

5 How does this prove Pythagoras' theorem? (The theorem states, "In any right triangle, the area of the square on the hypotenuse is equal to the sum of the areas of the squares on the remaining two sides.")

ATL2

Pairs

6 With a partner, look at each other's proofs. Take turns giving feedback by first asking questions to clarify anything that isn't quite clear. Then, give at least one positive comment. Follow that with a suggestion for improvement, if appropriate.

Reflect and discuss 5

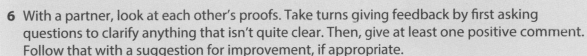

In small groups, discuss the following questions.

- What do the different ways of showing Pythagoras' theorem all have in common? Explain.

- Look on the internet for a video showing the water proof of Pythagoras' theorem.

- Which of the proofs that you have seen gives the clearest evidence? Explain your reasoning.

Applying the Pythagorean theorem

Pythagoras' theorem generalizes the relationship between the three sides of a right-angled triangle. It allows you to find the length of a side in any right-angled triangle provided you know the lengths of the other two sides. This is especially useful in situations where two of the sides are easy to measure, but the third is not.

Example 1

Q Measuring the height of very tall objects has been made easier by the invention of a laser rangefinder. The tallest tree in the world, named Hyperion, is a coastal redwood tree in California, USA. A surveyor standing 40 m away from the base of the tree uses a laser rangefinder to measure its height. He finds that the height of the tree is 116 m. How long is the laser beam from the surveyor to the top of the tree?

A

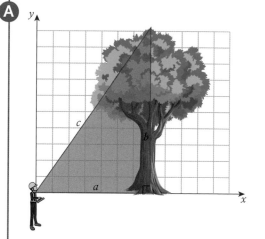

Construct a right-angled triangle. Be sure to indicate the right angle.
Label the legs a and b, and the hypotenuse c.

$$c^2 = a^2 + b^2$$
$$c^2 = 40^2 + 116^2$$
$$c^2 = 1600 + 13\,456$$
$$c^2 = 15\,056$$
$$c = 123 \text{ (3 s.f.)}$$

The length of the laser is 123 m.

Apply Pythagoras' theorem, using the values that you know: $a = 40$ m, $b = 116$ m.

Do not forget to take the square root in order to find the value of c.

Practice 1

1 Find the value of x in each diagram. Round your answers to the nearest tenth where necessary.

a
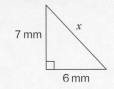
7 mm
x
6 mm

b
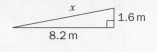
x
8.2 m
1.6 m

c

66 cm
x
40 cm

d
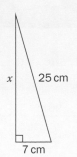
x 25 cm
7 cm

e
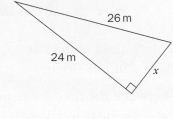
26 m
24 m
x

f
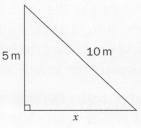
5 m   10 m
x

2 Use Pythagoras' theorem to determine whether the following triangles contain right angles.

a
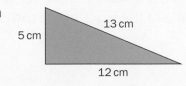
5 cm
13 cm
12 cm

b
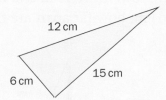
12 cm
6 cm   15 cm

c
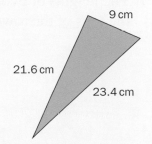
9 cm
21.6 cm
23.4 cm

3 The length of the diagonal of a rectangle is 50 cm. If the length of one side of the rectangle is 30 cm, calculate:

a the length of the other side

b the area of the rectangle.

4 The diagonals in a square are 10 cm long. Calculate the length of the sides of the square.

5 In the construction industry, one third of all accidents happen while using a ladder. To try to solve the problem, regulations have been introduced that specify how far the base of a ladder should be from the wall it is leaning against.

a If a ladder is to reach a height of 5.6 m, the base must be at least 1.4 m from the wall. How long should the ladder be in order to meet this safety regulation?

b A ladder measuring 4 m needs to have its base placed 97 cm from the wall. How high up the wall can the ladder reach?

▶ Continued on next page

6 Surveying is the science of constructing maps based on measurements. It is a process that cartographers use to calculate heights and distances between different points. A surveyor looks through a theodolite (a measuring device based on a telescope) at a measuring stick that is a known distance away. The telescope's line of sight and the measuring stick form a right angle. Since the surveyor knows the height of the stick and the horizontal distance to it, she can calculate the length and steepness of a hill or slope.

Suppose the horizontal distance between the theodolite and the top of the stick is 90 m and the stick measures 1.5 m. How far is the top of the theodolite from the bottom of the stick? Show your working.

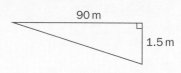

7 The heights of two vertical towers are 75 m and 65 m. If the shortest distance between the tops of the towers is 61 m, how far apart are the middle of the bases of the towers?

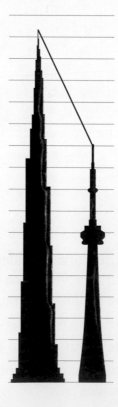

▶ Continued on next page

8 Over the last few decades, television technology has improved, allowing a much wider range of screen sizes. To enable television screen sizes to be easily compared, they are measured using the diagonal of the television.

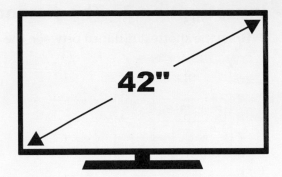

Suppose you have bought a 42-inch television and the cabinet you want to put it in has a height of 13 inches and a width of 39 inches. Will the television fit in the cabinet?

9 Groups of three whole numbers that satisfy Pythagoras' theorem are called "Pythagorean triples". Find the third number that would make these into Pythagorean triples. They are given in numerical order.

  a 3, 4, …

  b 5, 12, …

  c 8, …, 17

  d …, 40, 41

10 You proved Pythagoras' theorem by constructing squares on each of the sides. Does the formula work with other shapes, for example circles?

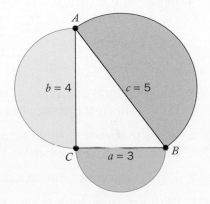

If semicircles are constructed on the sides of a right triangle whose sides measure 3 cm, 4 cm and 5 cm, show that Pythagoras' theorem still applies.

The Pythagorean theorem is so famous, it has been quoted in several movies! Unfortunately, not all of them have used the theorem correctly. If you search for "Scarecrow Pythagorean Theorem" on teachertube.com, you will see one of the more famous blunders!

The Pythagorean theorem is a very important tool in the study of triangles. However, it has also formed the basis of solutions in coordinate geometry as well.

Putting Pythagoras on the map

The shortest distance between any two points on the Cartesian coordinate plane can always be calculated using Pythagoras' theorem.

Investigation 2 – The distance formula

1 Draw the shortest distance between the coordinates below. Measure this distance.

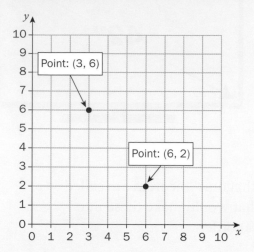

2 Create a right triangle with the segment you drew in step 1 as the hypotenuse.

3 Show how can you use Pythagoras' theorem to calculate the distance you measured.

4 Repeat the activity with the pairs of coordinates below. Plot each pair of points and create a triangle to illustrate your findings.

    a (2, 5) and (5, 1)   **b** (−13, 2) and (−1, 7)

    c (−1, −2) and (−7, −8)   **d** (15, −1) and (3, −6)

5 Using general points (x_1, y_1) and (x_2, y_2), generalize your findings to find a formula for the shortest distance between any two points.

6 Verify that your formula works for two more pairs of points. Show that you obtain the same result as when you plot the points and create a right triangle.

7 Justify why your formula works in every case.

8 With a partner, look at each other's distance formulae. Take turns giving feedback by first asking questions to clarify anything that isn't quite clear. Then, give at least one positive comment. Follow that with a suggestion for improvement, if appropriate.

Pairs

Reflect and discuss 6

- Does it matter which point you use first in your distance formula? Explain.

- Explain how the distance formula is an application of Pythagoras' theorem.

- Do you need to memorize this formula? Explain.

Example 2

Q When boating, distances are typically measured in nautical miles and speeds in knots (nautical miles per hour). Because one nautical mile is equal to one minute of latitude, this solves the problem of having to convert distances to latitude and longitude when navigating on open water.

Using the map below, calculate the shortest distance between the boat and Robin Hood's Bay. (Each distance is measured in nautical miles.)

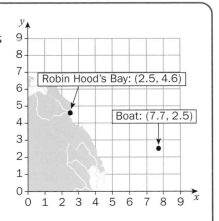

A

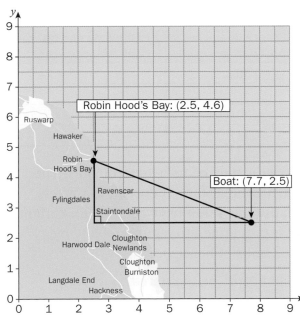

Construct a right-angled triangle. Be sure to indicate the right angle.

$d^2 = (x_2 - x_1)^2 + (y_2 - y_1)^2$
$d^2 = (7.7 - 2.5)^2 + (2.5 - 4.6)^2$
$d^2 = (5.2)^2 + (-2.1)^2$

$d^2 = 27.04 + 4.41$
$d^2 = 31.45$

distance $= \sqrt{31.45}$

distance $= 5.61$ nautical miles

Let Robin Hood's Bay with the coordinates (2.5, 4.6) be (x_1, y_1) and let the boat with the coordinates (7.7, 2.5) be (x_2, y_2).

Do not forget to take the square root to find the distance. Because you are dealing with distances, only the positive root makes sense.

The shortest distance between the boat and Robin Hood's Bay is 5.61 nautical miles.

Practice 2

1 Find the shortest distance between each of the following pairs of points. Show your working and round your answers to the nearest tenth.

    a $(-1, 4)$ and $(3, 10)$ **b** $(-7, -11)$ and $(4, -25)$

    c $(0, -6)$ and $(4, 3)$ **d** $(5, -2)$ and $(13, -2)$

2 When data is sent electronically, for example when you download a song, you don't want parts of it to be missing. Computer programmers, who imagine data as points in space, use Pythagoras' theorem to make sure that the data is where it should be.

Given an original point and its final location, determine whether or not a piece of data is where it should be (a distance d from its original location). Copy and complete this table.

Original location	Final location	Actual distance traveled	Distance data was supposed to travel (d)	Did the data end up where it was supposed to? (Y/N)
(2, 6)	(8, 14)		10	
(−1, 3)	(2, −1)		6	
(1, −1)	(−4, 5)		7	
(0, 7)	(−5, −5)		13	

3 In video games, the location of each character is tracked using coordinates. The program uses Pythagoras' theorem to calculate the distance between characters. It then compares this value with minimum distances that have been set for certain actions. For example, a role-playing game is programed to register a "hit" by a character's sword if the character is within 2 units of her opponent.

    a In which of the following scenarios would a "hit" be registered?

       i Player A (1, 4) and Player B (2, 2)

      ii Player C (−1, 2) and Player D (0, 1)

     iii Player E (3, 2) and Player F (4, 1)

     iv Player G (−2, −3) and Player H (−1, 0)

    b Establish a location for two players (other than those above) where a "hit" would be registered. Justify your answer.

4 Archeologists use a coordinate grid system when excavating a site. This allows them to easily record the location of artefacts at a site, which simplifies future excavations.

▶ Continued on next page

If the coordinates (in cm) of the top and bottom of the arm bone of this dinosaur skeleton are (6.2, 22.3) and (9.1, 11.8), find the length of the arm bone, rounded to the nearest tenth of a centimetre. (Assume that the arm bone is perfectly straight.)

5 The distance between (–2, 6) and (4, y) is 10 units. Find the values of y. Show your working.

6 A ship is transporting goods from China to the United States. It leaves Shanghai, whose coordinates in nautical miles are (–2140, 1320). It stops off in Honolulu (2152, 1470) to refuel and then continues on to Los Angeles (5120, 2016). Find the total length of the voyage in nautical miles.

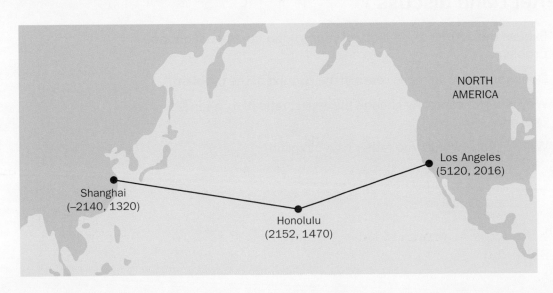

Formative assessment – How can "the same" be "not equal"?

The screen size of cell phones, like televisions, is the measure of their diagonal. Do cell phones advertised with the same screen size also have the same area?

The *aspect ratio* of a screen is the ratio of the rectangular side lengths. Televisions and cell phones typically have an aspect ratio of 16 : 9. (Their sides are in a ratio of 16 : 9, but they may be larger or smaller than those values.) However, some newer models have aspect ratios of 18.5 : 9.

The screen size of two phones is given as 127 mm, but their aspect ratio is different.

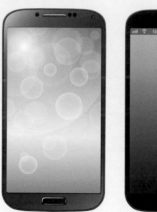

a Find the side lengths of a rectangular cell phone screen with an aspect ratio of 16 : 9 whose diagonal measures 127 mm.

b Find the area of this screen. Show all of your working and round your answer to the nearest hundredth.

c Repeat these calculations for a cell phone with the same diagonal measurement, but with an aspect ratio of 18.5 : 9.

d Which phone's screen has a larger area?

Reflect and discuss 7

- Explain the degree of accuracy of your solution to the cell phone problem.

- Does your solution make sense in the context of the problem?

- Why would a company change the aspect ratio of its phone? Explain.

- Which phone would you rather have? Explain.

Relationships between triangles

Pythagoras' theorem establishes the relationship between the lengths of the sides in a right triangle. While this is an important relationship, it only applies to right triangles. Are there other relationships between the side lengths in right triangles? Do relationships exist between side lengths in other types of triangle? Are there relationships between different triangles?

Similar and congruent triangles

Regardless of the type of triangle, two triangles can be *congruent* to each other or *similar* to each other or neither.

Activity 5 – Congruent triangles

Below are some pairs of *congruent* triangles.

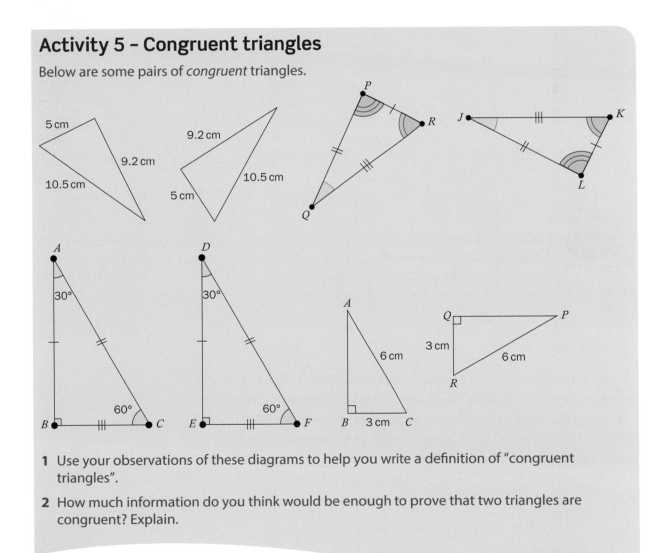

1 Use your observations of these diagrams to help you write a definition of "congruent triangles".

2 How much information do you think would be enough to prove that two triangles are congruent? Explain.

▶ Continued on next page

3 Do you think the following pairs of triangles are necessarily congruent? Justify your answer.

a

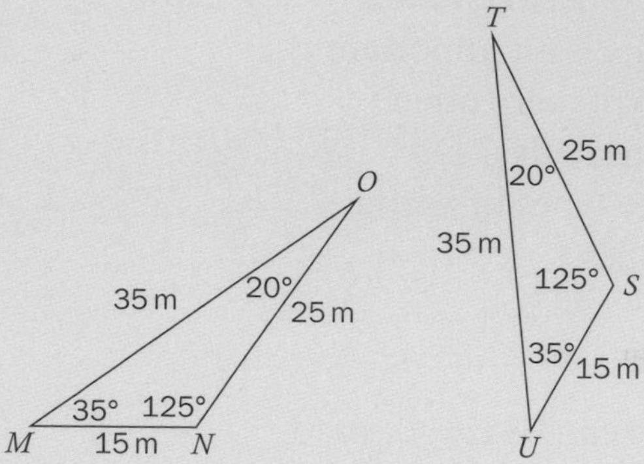

b

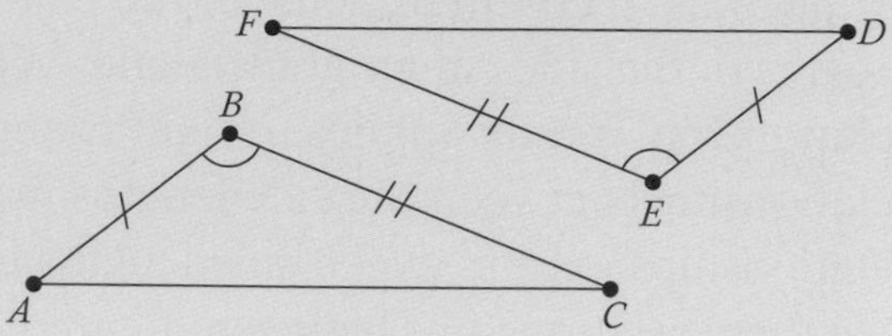

c

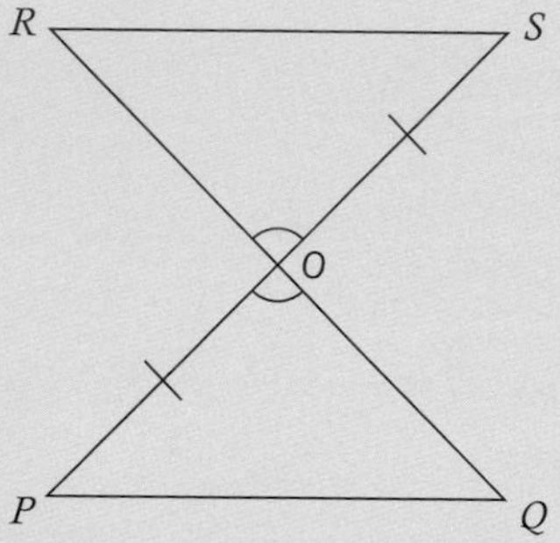

d

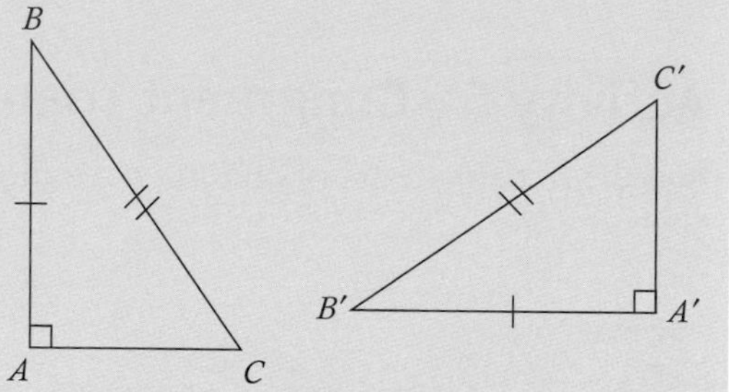

WEBLINK

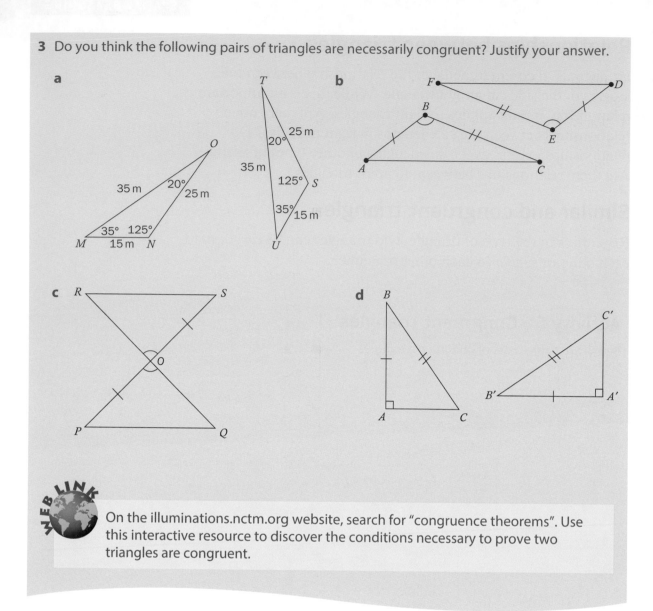

On the illuminations.nctm.org website, search for "congruence theorems". Use this interactive resource to discover the conditions necessary to prove two triangles are congruent.

Similar triangles also have a particular relationship to each other, although this is different than that of congruent triangles. Knowing that two triangles are similar allows you to develop further mathematical principles as well as solutions to a wide range of problems.

Investigation 3 – Properties of similar triangles

You can draw the triangles in this investigation by hand or you can use dynamic geometry software to achieve the same results.

criterion B

1 Draw two triangles of different sizes that both have a 50° angle and a 70° angle.

2 Measure the lengths of the sides. What do you notice about the lengths of the sides of the two triangles?

3 Compare your results with those of a peer. What do you notice?

4 Draw two right triangles of different sizes that have the same angles and measure the three sides.

5 Do your results from steps 2 and 3 still hold?

6 In a table like the following, generalize the relationships between measures in congruent and similar triangles.

	Relationship between angles	Relationship between sides
Similar triangles		
Congruent triangles		

ATL1

7 Verify the relationship between the sides of similar triangles for another pair of triangles.

8 Justify why the relationship between the sides of similar triangles exists.

Reflect and discuss 8

- Are all right triangles similar? Explain.

- Can two triangles have the same angle measures but side lengths that are not proportional? Explain.

Proving triangle similarity

The symbol for "similar" is ~. The notation for showing that two triangles are similar is

$$\Delta ABC \sim \Delta DEF$$

The corresponding angles and sides must be written in the same order in each triangle.

Reflect and discuss 8

If $\triangle ABC \sim \triangle DEF$,

- write down the angles that are equal
- write down the ratios of the sides that are proportional.

> Your ratios should be in the form "$\dfrac{JK}{MN}$".

In Investigation 3, you developed the *principle of similarity* between two triangles. If two triangles are similar, then the following are true:

- their corresponding angles are equal
- their corresponding sides are proportional.

In order to prove that two triangles are similar, you need to show that one of these conditions is true. Once one of them is proved to be true, then you know that the other statement is also true.

ATL2

Activity 6 – Similarity postulates

Pairs

Perform this activity with a peer, taking turns to describe your thinking. Also take turns giving feedback by first asking questions to clarify anything that isn't quite clear. Then, give at least one positive comment. Follow that with a suggestion for improvement, if appropriate.

1 Each diagram shows a pair of similar triangles. Identify the triangles that are similar (e.g. $\triangle ABC \sim \triangle DEF$). Which principle of similarity allows you to prove them to be similar? Justify your answer.

a

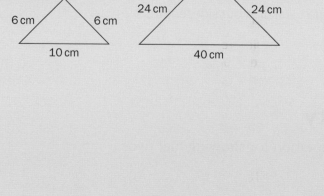

b

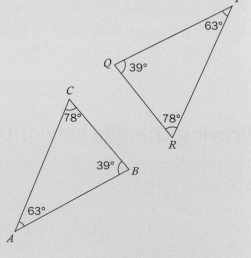

▶ Continued on next page

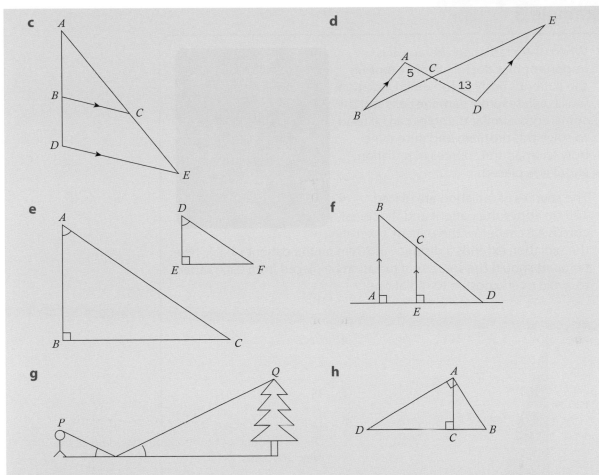

2 One of the postulates used to prove that two triangles are similar is called "AA" for "Angle-Angle". It states, "If two angles of one triangle are congruent to two angles of another triangle, then the two triangles are similar." Explain why you only need to prove that two angles are congruent instead of all three.

3 Another similarity postulate is "SSS". What do you think this refers to? Use an example to show how it proves that two triangles are similar.

4 A third postulate is "SAS" which states, "If two sides of one triangle are proportional to two sides of another triangle and the included angle in both triangles is congruent, then the triangles are similar." Draw an example of two similar triangles and indicate the measures of two sides and the included angle to demonstrate what this postulate means.

Applications of similar triangles

Once you have established that two triangles are similar, you know that their angles are congruent and that their sides are proportional. This information can help you to solve a wide range of problems, some of which are more critical than you might think.

Example 3

Q When using radiation therapy, it is important that doctors avoid exposing the patient's body to too much radiation, which can severely damage cells and the spinal cord. Similar triangles can be used to solve this problem and determine how far apart the sources of radiation should be placed.

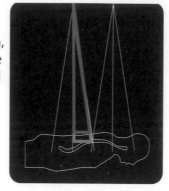

The sources of radiation are placed 120 cm above the patient and the spinal cord is 4.8 cm below the patient's skin. If the radiation extends a distance of 20 cm on the patient's back, how far apart should the sources of radiation be placed from each other to avoid overexposure to radiation?

A

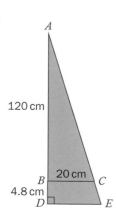

> Draw a diagram of the situation, including all the measurements.

The two triangles are similar by the AA postulate. They each have a right angle and angle A is the same in each.

> Establish that the triangles are, in fact, similar.

$\triangle ABC \sim \triangle ADE$

$$\frac{AB}{AD} = \frac{BC}{DE} = \frac{AC}{AE}$$

> Because the triangles are similar, their sides are proportional. Write down the ratios that are equivalent.

$$\frac{120}{124.8} = \frac{20}{DE}$$

> Fill in the measures that you have and solve.

$120DE = 20(124.8)$

$120DE = 2496$

$DE = 20.8\,cm$

$2(20.8) = 41.6\,cm$

> Each large triangle has a base of 20.8 cm, which means the sources of radiation need to be 41.6 cm away from each other.

The radiation sources should be placed at least 41.6 cm away from each other to avoid radiation overexposure.

Practice 3

1 Indicate which of the following are pairs of similar triangles and name the postulate that you used to establish similarity.

a

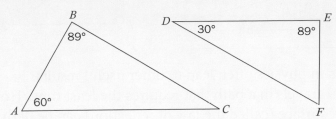

b

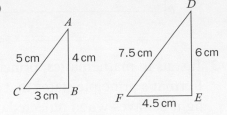

c

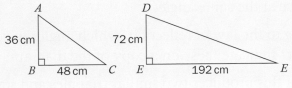

d

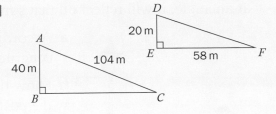

e

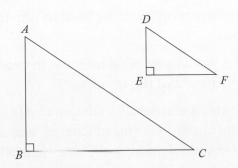

f

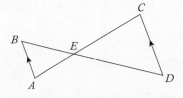

2 Indicate which pairs of triangles in each diagram are similar and write down the similarity postulate you used to establish the relationship. Then find the missing measurement. Show your working and round your answers to the nearest hundredth where appropriate.

a

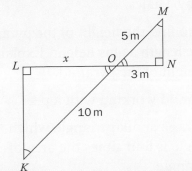

b

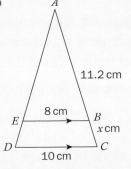

c

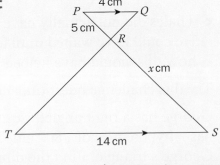

▶ Continued on next page

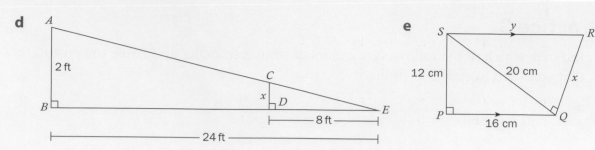

3 Just as in mathematics, principles in physics often lead to other useful results. Fermat's principle states that light travels in a path that requires the least time. How light behaves when reflected off a surface (called the law of reflection) can be derived from this principle. The law of reflection states that when light hits a reflective surface at an angle, it will reflect off that surface at the same angle.

a According to the law of reflection, which angles will be congruent? Indicate these on a copy of the diagram.

b Show that this produces two similar triangles and state the postulate you used.

c Describe how this information could be used to find the height of the lamppost.

d What measurements would you need to know in order to actually calculate the height of the lamppost?

4 Thales was a Greek mathematician, philosopher and a teacher of Pythagoras. He supposedly asked the Egyptian priests how tall the Great Pyramid of Cheops was and, when they refused to tell him, he set out to solve the problem himself. One method that he is said to have used involved measuring shadows. Thales, who was 176 cm tall, measured his shadow to be 160 cm. At exactly the same time, he measured the shadow of the pyramid to be 13 320 cm.

a Draw a diagram of this situation. Indicate the similar triangles and justify how you know that the triangles are similar.

b Find the height of the Great Pyramid as calculated by Thales.

c Thales did not actually have to use similar triangles to find the height of the pyramid. He could have waited until his shadow was the same length as his height. Explain how this would have helped him to determine the height of the pyramid.

Pairs

5 Try this challenge question, and compare your answer and working with a peer.

George has a right-angled isosceles triangle and Prita has a similar triangle which has a hypotenuse of $\sqrt{800}$ cm. The area of Prita's triangle is four times the area of George's triangle. Find the dimensions of George's triangle.

Trigonometric ratios

Understanding the relationships between the sides in right-angled triangles is very useful in mathematics. You have already established Pythagoras' theorem and the principle of similarity. Relationships also exist between a right triangle's sides **and** its angles. This is where the principles of trigonometry begin.

The sine ratio

In triangles, angles are generally named using upper case letters while sides are named using lower case letters. A side is labeled with the same letter as the angle that is across from it, as shown in the diagram.

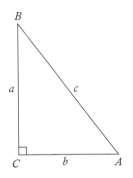

Understanding the relationships between the sides in right-angled triangles is a very important technique and leads to developing even more skills. You have already established Pythagoras' theorem and the principle of similarity with right triangles.

Activity 7 – The sine ratio

1 Using a protractor and a ruler, draw two right-angled triangles of different sizes but both having one of the acute angles measuring 60 degrees.

2 Measure the lengths of the hypotenuse and the side opposite the 60° angle in your triangles. Record your results in the first two rows of a table like this one.

Triangle	Measure of opposite side (O)	Measure of hypotenuse (H)	O + H	O − H	OH	$\dfrac{O}{H}$
1						
2						
3						
4						

Pairs

3 Share your results with a peer so that you have data for four right triangles.

4 Explore the relationships between the hypotenuse and the opposite side by carrying out the different calculations. Add your results to the table.

5 Examine your results. Which calculations produce observable patterns? What patterns do you observe?

6 Draw two right triangles each of which has a 45° angle. Then repeat steps 1 through 4.

▶ Continued on next page

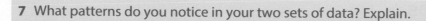

7   What patterns do you notice in your two sets of data? Explain.

ATL2

Pairs

8   With a partner, discuss the patterns you both found. Take turns giving feedback by first asking questions to clarify anything that isn't quite clear. Then, give at least one positive comment. Follow that with a suggestion for improvement, if appropriate.

Reflect and discuss 9

- Explain why the value of the ratio $\dfrac{O}{H}$ is always between 0 and 1.
- What mathematical principle underlies the reason why the ratio of $\dfrac{O}{H}$ is the same in some of your triangles? Explain.

Did you know...?

Thankfully, you do not have to draw triangles in order to calculate the sine ratio for an angle. Technology can be used instead, although the technology has changed dramatically over the years.

The *slide rule* (pictured right) was invented in the 17th century. This tool could perform calculations involving many mathematical operations, including finding the sine ratio.

In the 1970s, scientific calculators replaced the slide rule. They were more powerful and much more user-friendly. Calculators use an *algorithm* to find the sine ratio.

You can use the fact that the sine ratio is the same in similar right triangles to calculate missing sides and/or angles.

Example 4

Q Benjamin Franklin used a kite and a key to discover an important scientific principle: lightning is a form of electricity. Contrary to popular belief, it is unlikely that his kite was actually struck by lightning, but that wasn't necessary for the experiment. All Franklin needed was to get his kite high in the air during a thunderstorm.

Suppose Franklin, whose height was 1.75 m, attached 16 m of string to the kite and it flew at an angle of 70° above the horizontal. Assuming the string was completely taut, how high off the ground was the kite? Round your answer to the nearest hundredth.

▶ Continued on next page

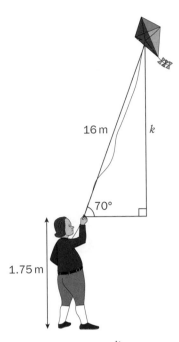

Construct a right-angled triangle. Be sure to indicate the right angle.

Label the measurements that you know: the angle of the kite string, the length of the string and Benjamin Franklin's height.

Label the side that you want to find with a letter.

16 m

k

70°

1.75 m

$\sin 70° = \dfrac{\text{opposite}}{\text{hypotenuse}}$

$\sin 70° = \dfrac{k}{16}$ Write the sine ratio with the information that you know.

$16 \sin 70° = k$ Isolate k.

$15.04 = k$ Use your calculator to find the value of k.

Make sure your calculator is in degree mode!

$15.04\,\text{m} + 1.75\,\text{m} = 16.79\,\text{m}$ Do not forget to add on Franklin's height.

The height of the kite off the ground is 16.8 m to 3 s.f.

On some calculators, you enter the number of degrees first and then press "sin" to find the sine ratio. On others, you type "sin 70" like you would write it in your notebook.

The sine ratio relates the length of the hypotenuse and the length of the side opposite a given angle. You can use the sine ratio to find the length of either of these sides as long as you know the length of the other one and the size of the angle. Is it possible to find the size of an angle given the lengths of these two sides instead?

Example 5

Q Planes use principles of physics to achieve flight. However, actually flying the plane involves a considerable amount of mathematics. Pilots can approximate some of the calculations while computers do most of them.

A plane is flying at an altitude of 10 000 m. It begins its descent when it is approximately 191 000 m from the runway. What is the plane's angle of descent?

A

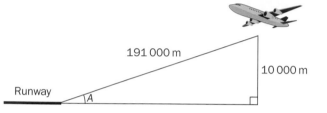

Construct a right-angled triangle. Be sure to indicate the right angle. Label the height of the plane and its distance from the runway. Indicate the angle to be found.

$$\sin A = \frac{\text{opposite}}{\text{hypotenuse}}$$

$$\sin A = \frac{10\ 000}{191\ 000}$$

Write the sine ratio with the information that you know.

$$A = \sin^{-1}\left(\frac{10\ 000}{191\ 000}\right)$$

$$A = 3.0°$$

In order to find the angle when you know the ratio, use the arcsin, or $\sin^{-1}$, command on your calculator.

The plane's angle of descent is 3.0° above the horizontal.

Practice 4

1 Use your calculator to find the given ratio. Round your answers to the nearest hundredth.

    a $\sin 58°$ **b** $\sin 12°$ **c** $\sin 74°$

    d $\sin 30°$ **e** $\sin 45°$ **f** $\sin 62°$

▶ Continued on next page

2 Use your calculator to find the size of the angle with the given sine ratio. Round your answers to the nearest tenth of a degree.

a $\sin A = \dfrac{1}{2}$ **b** $\sin A = \dfrac{3}{4}$ **c** $\sin A = \dfrac{12}{23}$

d $\sin A = \dfrac{6}{11}$ **e** $\sin A = 0.64$ **f** $\sin A = 0.15$

3 Find the ratios $\sin B$ and $\sin C$ in each of the following triangles. Simplify all fractions.

a
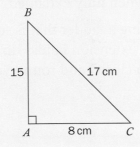

b
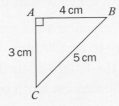

c
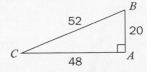

d
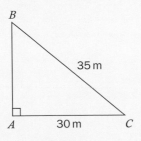

e

f
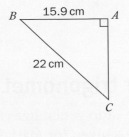

4 Find the missing measurements. Round your answers to the nearest tenth.

a

b
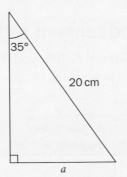

c
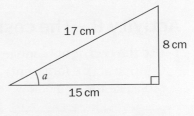

d
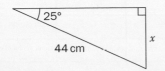

e
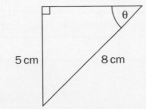

f
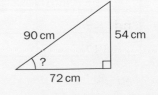

▶ Continued on next page

5 a Mesfin bought a 6-meter ladder in order to paint some murals at school. Ladder safety regulations state that the ladder should be placed at an angle of no more than 75 degrees with the ground. What is the maximum height that Mesfin will be able to reach with his ladder?

b Extension ladders were developed to solve the need for ladders that are long enough to reach high places but compact enough to store anywhere. These ladders can reach a range of heights. When closed, they are half their length when fully extended.

If you want to reach a height of 4.5 m using a ladder that is set up according to regulations, what length extension ladder do you need? Give the length of the ladder when closed and round your answer to the nearest tenth of a meter.

Other trigonometric ratios

The sine ratio generalizes the relationship between the side opposite a given (or marked) angle and the hypotenuse. Are there other ratios between pairs of sides?

Activity 8 – The cosine and tangent ratios

1 Use the two right triangles with a 60° angle that you constructed in Activity 8. Label the sides, O (opposite), A (adjacent) and H (hypotenuse). Measure and record these side lengths in a table like the one below.

Triangle	Length of opposite side (O)	Length of hypotenuse (H)	Length of adjacent (A)	$\dfrac{A}{H}$	$\dfrac{O}{A}$
1					
2					
3					
4					

▶ Continued on next page

2 Share your results with a peer so that you have data for four right triangles, each with a 60° angle.

3 Calculate the ratios in the last two columns. What do you notice?

4 Repeat steps 1 through 4 for right triangles containing a 45° angle. You may use the ones you drew in Activity 8.

5 What patterns do you notice between your two sets of data? Explain.

6 Research the names of these two ratios.

Reflect and discuss 10

- Explain why the value of $\frac{A}{H}$ is always between 0 and 1.
- Describe any limitations on the value of $\frac{O}{A}$.
- What mathematical principle underlies the reason why these ratios are the same in certain right triangles? Explain.

The three trigonometric ratios relate the side lengths in any right triangle, where angle A is not the right angle.

$$\sin A = \frac{\text{opposite}}{\text{hypotenuse}}$$

$$\cos A = \frac{\text{adjacent}}{\text{hypotenuse}}$$

$$\tan A = \frac{\text{opposite}}{\text{adjacent}}$$

You can solve problems with all these generalized relationships. If you know two measures in a right triangle, you can use the relevant ratio to solve for a third measurement.

Some people use the acronym SOHCAHTOA to remember the three fundamental trigonometric ratios.

$$s = \frac{o}{h} \quad \text{SOH}$$

$$c = \frac{a}{h} \quad \text{CAH}$$

$$t = \frac{o}{a} \quad \text{TOA}$$

Feel free to come up with your own memory aid!

Example 6

Q In the past, it was extremely difficult for people in a wheelchair or pushing a stroller to access buildings with steps up to the entrance. To solve this problem, wheelchair ramps were developed. Many countries now have strict guidelines for wheelchair ramps so that the angle of the ramp is not too steep.

The most common regulation states that for every 12 meters of horizontal distance the ramp can rise a maximum of 1 meter. What is the maximum angle of a wheelchair ramp?

A

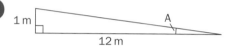

$$\tan A = \frac{\text{opposite}}{\text{adjacent}}$$

Sketch a right-angled triangle. Be sure to indicate the right angle. Label the lengths of the sides. Indicate the angle to be found.

You have the lengths of the opposite and the adjacent sides to the angle you want to find, so use the tangent ratio.

$$\tan A = \frac{1}{12}$$

Write the tangent ratio with the information that you know.

$$\tan^{-1}(\tan A) = \tan^{-1}\left(\frac{1}{12}\right)$$

$$A = \tan^{-1}\left(\frac{1}{12}\right)$$

Remember that solving equations involves performing the same operation on both sides. The opposite of "tangent" is "arctan". On a calculator arctan is denoted as $\tan^{-1}$.

$$A = 4.8°$$

Make sure your calculator is in degree mode.

The maximum angle of a wheelchair ramp is 4.8°.

Measuring the lengths of sides of a right triangle seems simple enough to do. However, how do you measure an angle? In geometry class, you might have used a protractor. Land surveyors use a transit level or a theodolite to measure both horizontal and vertical angles. Forestry professionals use a tool called a *clinometer* to measure angles in order to calculate the height of a tree. You can build your own clinometer, which you will use in the summative assessment task for this unit.

Activity 9 – Clinometer

You can download a free clinometer app instead of making the one in this activity. Using this app, you view the desired object along the edge of your phone and read the angle on the screen.

To make your own clinometer, you will need the following materials: cardboard, string, a straw, a small weight and a photocopy of a protractor.

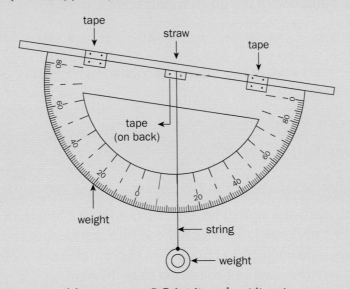

1. Search online for a "printable protractor". Print it and cut it out.

2. Glue the protractor onto the cardboard and then cut around it so there is no excess cardboard.

3. Pierce a hole through the protractor and cardboard at the midpoint of the line connecting 0° and 180°.

4. Thread the string through the hole and tie a knot in it or stick it in place with tape. Make sure some string hangs below the protractor when you hold it with the horizontal edge at the top.

5. Tie the weight to the end of the string.

6. Tape the straw to the protractor along the line between 0° and 180°. It will cover the hole you made previously.

7. To use your clinometer, hold it so that the horizontal edge is at the top and the curved edge is at the bottom. The weight should hang vertically down. Look through the straw at the top of the object whose height you are trying to calculate.

8. Read the angle on the clinometer. To obtain the angle of the object, find the difference between your angle and 90° – take the absolute value of this difference.

9. Measure how far your feet are from the edge of the object. Draw a diagram and use trigonometry to find the height of the object. Remember to add on your own height up to your eye level.

10. Compare your value with that of a peer, especially if he/she used the clinometer app.

Practice 5

1 Find the value of each trigonometric ratio. Round your answers to the nearest thousandth.

   a $\cos 62°$ **b** $\tan 17°$ **c** $\sin 88°$ **d** $\tan 34°$ **e** $\cos 29°$ **f** $\sin 44°$

2 Find the size of the angle with the given trigonometric ratio.

   a $\tan A = \dfrac{2}{3}$ **b** $\cos A = \dfrac{5}{6}$ **c** $\cos A = \dfrac{9}{11}$ **d** $\tan A = \dfrac{20}{7}$ **e** $\cos A = 0.41$ **f** $\tan A = 3.42$

3 a For this triangle, find each of the following trigonometric ratios. Simplify your fractions where necessary.

    $\sin B$ $\cos A$ $\tan B$

    $\sin A$ $\cos B$ $\tan C$

  b Explain why $\cos B = \sin C$ and $\sin B = \cos C$.

4 A ladder is leaning against a wall. The ladder has a known length of 3 m. Cassandra wishes to find the vertical height to which the ladder reaches but she only has a 15 cm ruler. She measures a distance of 15 cm up the ladder and drops a piece of string to the ground. She measures the length of the string to be 10.6 cm.

  a From this information, determine the vertical height to which the ladder reaches. Show your working.

  b Did you use similar triangles or trigonometry? Use the method you didn't use in step **a** to verify your answer.

5 Janine is going to use her clinometer to measure the height of the tree near her house. She is worried that it may hit her home if it falls in a big storm, and has measured that its base is 18 m away from her house. Janine is 1.80 m tall. Her eyes are 10 cm from the top of her head. The horizontal distance between Janine's eyes and the tree is 20 meters. The clinometer measures the angle from Janine's eye level to the top of the tree to be 40 degrees.

  a Find the height of the tree.

  b Should Janine be worried about the tree hitting her house? Justify your answer.

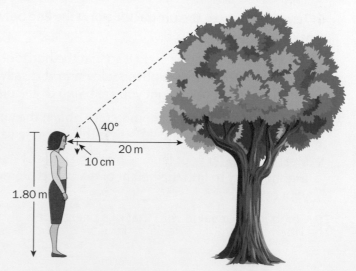

▶ Continued on next page

6 The angle of repose is the steepest angle at which dry, loose material is stable when put in a pile. When the angle of repose is exceeded, the material will slide. In snowy mountains, this can result in an avalanche. For a science experiment, you create a conical pile of sand that is as steep as you can make it. The pile is 11 cm high and has a radius of 16 cm. What is the angle of repose of this sand?

7 In physics, when an object travels at an angle, it is easier to break up the motion into a horizontal component and a vertical component. The principles of physics can then be applied to both directions. Suppose an object travels with a velocity (speed) of 25 m/s at an angle of 40 degrees with the ground. Find the horizontal and vertical components of the object's velocity.

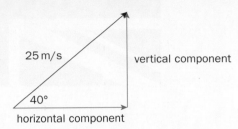

25 m/s
vertical component
40°
horizontal component

Formative assessment – The creation of flags

criterion **C,D**

Do you know the story behind your country's flag? A flag can reveal much about a country's history and culture. How are flags designed? Is this a creative process or a scientific one or maybe a combination of both? In this task, you will analyse the Union Jack (the flag of the United Kingdom) before attempting to design one of your own.

The Union Jack

How do you solve the problem of creating a flag for three countries that already have their own individual flags? Add them all together! The Union Jack combines the flags of three of the countries that came together to form the United Kingdom: the English Cross of St George, the Scottish Cross of St Andrew and the Irish Cross of St Patrick. (The Welsh flag was not included, as Wales was considered a principality rather than a separate kingdom.)

The Union Jack, with some of its dimensions, is shown on the next page.

▶ Continued on next page

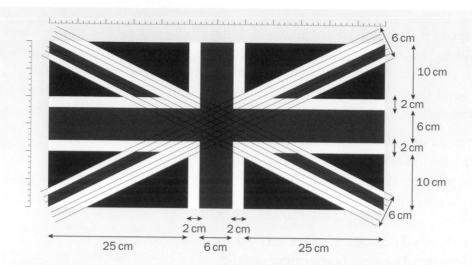

a Using your knowledge of similar triangles, Pythagoras' theorem and trigonometry, find the measurements of the blue triangles. Give side lengths and angles. Show your working.

Design your own flag

Did you know that the Olympic rings include colours from the flag of every country that participates?

b You will now design a flag of your own. Feel free to take inspiration from current flags that you have seen. Your flag must contain colours that are in the Olympic rings and it should have a height that is one half of its length. Use your knowledge of Pythagoras' theorem and trigonometry to achieve an aesthetically pleasing flag. Show all of the dimensions and all of your working.

c Explain the story behind your flag and why you have chosen the colours and elements of your flag.

ATL2

Pairs

d With a partner, look at each other's flag and the supporting work. Take turns giving feedback by first asking questions to clarify anything that isn't quite clear. Then, give at least one positive comment. Follow that with a suggestion for improvement, if appropriate.

Unit summary

Pythagoras' theorem states that

$$c^2 = a^2 + b^2$$

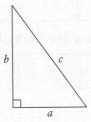

where *a* and *b* are the lengths of the *legs* of a right triangle and *c* is the length of the *hypotenuse*.

A triangle is said to be *congruent* to another triangle if one triangle is an exact copy of the other. Sides have exactly the same length and angles are exactly the same size.

A triangle is said to be *similar* to another triangle if one triangle is an exact enlargement of the other. The corresponding angles in the triangles must be exactly the same size. The sides of similar triangles are proportional to one another.

Similarity can be established using any one of the three *postulates*.

Postulates		Two triangles are similar if …
AA	Angle–Angle	two corresponding angles are congruent.
SAS	Side–Angle–Side	the lengths of two corresponding sides are proportional and the included angle is congruent.
SSS	Side–Side–Side	the lengths of all corresponding sides are proportional.

If $\triangle ABC \sim \triangle ADE$, then $\dfrac{AB}{AD} = \dfrac{BC}{DE} = \dfrac{AC}{AE}$.

The three fundamental trigonometric ratios establish relationships between the sides of a right triangle.

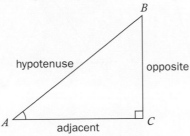

$$\sin A = \frac{\text{opposite}}{\text{hypotenuse}} \qquad \cos A = \frac{\text{adjacent}}{\text{hypotenuse}} \qquad \tan A = \frac{\text{opposite}}{\text{adjacent}}$$

Unit review

 criterion A

📖 **Launch additional digital resources for this chapter**

Key to Unit review question levels:

Level 1–2   **Level 3–4**   **Level 5–6**   **Level 7–8**

1 **Explain** how the diagram below can be used to prove Pythagoras' theorem.

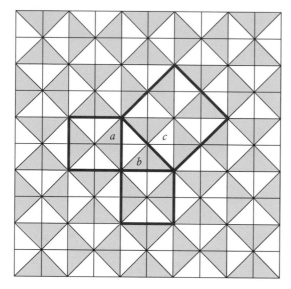

2 Are the following pairs of triangles similar? If they are, **write down** the similar triangles and the similarity postulate used.

a

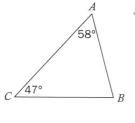

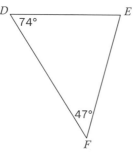

b

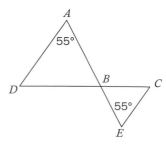

c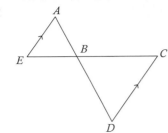

d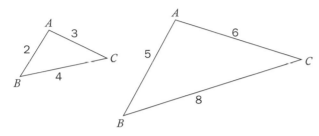

e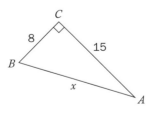

3. Find the value of the following trigonometric ratios. Round your answers to the nearest hundredth.

   a $\tan 50°$ **b** $\sin 38°$ **c** $\cos 75°$

   d $\sin 8°$ **e** $\cos 71°$ **f** $\tan 26°$

4. Find the measure of the angle with the given trigonometric ratio. Round your answers to the nearest tenth of a degree.

   a $\cos A = \dfrac{5}{8}$ **b** $\tan A = \dfrac{4}{3}$ **c** $\sin A = \dfrac{4}{9}$

   d $\tan A = 1.15$ **e** $\sin A = 0.89$ **f** $\cos A = 0.71$

5. Find the size of the unknown side in each triangle.

   a 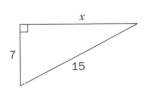   **b**

   c   **d**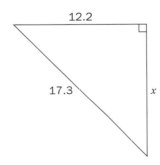

6 **Write down** the similarity postulate that proves the triangles in each pair are similar. Then find the missing measurement. Round your answers to 3 s.f.

a

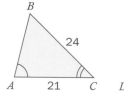

b

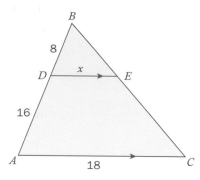

c

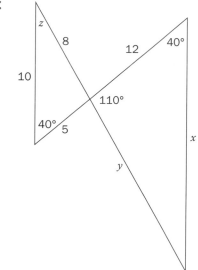

7 Find the value of the indicated measure in each of the following right triangles.

a

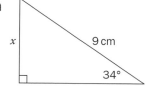

b

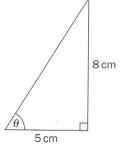

c

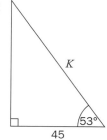

d

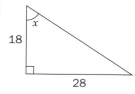

8 In ancient Egypt, when the Nile flooded, it often erased the boundaries between landowners. An accurate surveying process was necessary in order to re-establish property lines. Land surveyors used Pythagoras' theorem to verify that property lines were correctly drawn by calculating how long they should be.

Two farmers occupy a property that is in the shape of a rectangle. Their property line is the diagonal of the rectangle. Find the length of the property line if the sides of the property measure 48 m and 55 m.

9 Find the distance between each pair of points.

    a (0, –2) and (–3, 6)   **b** (–5, 1) and (3, –8)

    c (4, 4) and (–2, 8)   **d** (–10, –3) and (–7, 1)

10 Subduction is a process where a heavier oceanic plate plunges underneath a lighter continental plate. The process can be either deep or shallow, depending on the angle of subduction. In very old subduction zones, a mountain range, called a volcanic arc, is formed directly above the point where the heavier plate reaches a depth of 100 km. A deep trench along the boundary of the two plates is also formed.

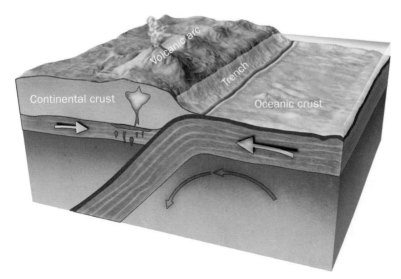

A volcanic arc is located a horizontal distance of 250 km from the trench.

    a **Draw** a diagram and indicate any relevant information.

    b Find the angle of subduction. Round your answer to the nearest tenth of a degree.

11 An isosceles triangle has a base of 6 cm and its equal sides are 5 cm long. **Calculate:**

   a the perpendicular height of the triangle

   b the area of the triangle.

12 The owners of a house want to replace the steps leading up to their back porch with a ramp. The porch is 1 meter off the ground, and to comply with building regulations, the ramp must make an angle of 4.8 degrees with the ground.

   a Find how long the slope of the ramp will be. **Show** your working.

   b Find the horizontal distance between the porch and the start of the ramp. **Show** your working.

13 Airport meteorologists keep an eye on the weather to ensure the safety of airplanes. One thing they watch is the cloud ceiling. The cloud ceiling is the lowest altitude at which solid cloud is visible. If the cloud ceiling is too low, the planes are not allowed to take off or land.

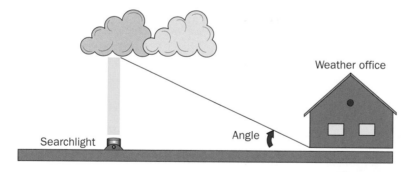

One process a meteorologist can use to find the cloud ceiling at night is to shine a searchlight that is located a fixed distance from her office vertically into the clouds. Then she measures the angle from the office to the spot of light on the cloud.

The searchlight is located 40 m from the office.

   a If the angle to the spot of light on the cloud is 38 degrees, how high is the cloud ceiling?

   b The minimum height of the cloud ceiling for planes to safely take off or land is 305 m. What is the angle to the spot of light on the cloud when the cloud ceiling is at the minimum height?

14 During a solar eclipse, the Moon passes between the Sun and Earth and covers the Sun almost entirely, despite being considerably smaller. How is this possible?

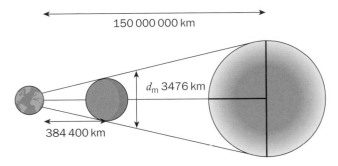

150 000 000 km

d_m 3476 km

384 400 km

a **Label** the triangles. **Write down** the triangles that are similar and indicate the similarity postulate that you used.

b On average, the Earth is 384 400 km from the Moon and 150 000 000 km from the Sun. The diameter of the Moon is 3476 km while the diameter of the Sun is 1 400 000 km. **Use** this information to determine why the Moon covers almost all of the Sun during an eclipse.

15 A television manufacturer made the following claim in an advertisement:

"Our 80-inch AQUOS TV delivers more than double the screen area of a 55-inch TV, for an amazing viewing experience."

a **Show** that an 80-inch television can deliver more than double the screen area of a 55 inch television.

b **Show** that this claim is true even if both televisions have an aspect ratio of 16 : 9. (Aspect ratio is length:width)

16 In order to describe the locations of celestial bodies, astronomers use a three-dimensional coordinate system with our Sun at the origin (0, 0, 0). All other objects are given a coordinate in the form (x, y, z).

a **Write down** the distance formula for a pair of 3-dimensional coordinates.

b Find the distance between our Sun and each of the following stars (distances given in light years). Round your answers to the nearest tenth.

   i Polaris (99.6, 28.2, 376.0)

   ii Alpha Centauri (−1.8, 0.0, 3.9)

   iii Sirius (−3.4, −3.1, 7.3)

c Find the distance from Polaris to Alpha Centauri.

Summative assessment

criteria **C, D**

Measuring the immeasurable

In this task, you will calculate the height of an object that cannot be measured easily. You will calculate its height using three different methods and then reflect on how accurate you think your answers are and the usefulness of the methods.

The object

Find an object of interest to you whose height would be difficult to measure directly. Your chosen object needs to have some flat, level ground around it and you must be able to access a point at the base of the object which is directly under its maximum height. This is so that you can measure the horizontal distance to the object.

Suitable objects might include a tree, a building, a flagpole, etc. (A mountain would not work since you can't be directly under its highest point.)

It is very important that the object is not on a sloping surface so that the points from which you take your measurements are all on the same level as the base of the object.

Method 1

a Standing away from the object, use a clinometer to measure the angle to the top of the object. Next measure the angle to the bottom of the object using the clinometer. Finally, measure how far away you are standing from the object.

b Draw a diagram to represent the object and the place where you took the measurement, and indicate the measurements on the diagram.

c Use trigonometry to calculate the height of the object.

Method 2

a Use the clinometer to measure the angle to the top of the object from a point away from its base.

b Move a few meters towards the object and measure the new angle to the top of the object. Be sure to measure the distance between the two points from which you took your measurements. **You are not allowed to measure/know how far you are from the object.**

c Draw a diagram to represent the object, the places where you took the measurements and the measurements themselves. Use trigonometry to calculate the height of the object. Remember to add the height of the clinometer to your result.

Method 3

Use a third method that applies similar triangles to *calculate* the height of the object. You may use one of the methods that has been described in this unit.

Reflect and discuss

- Describe whether or not your answers make sense.
- Explain the accuracy of your results. Did all methods give similar answers? How close do you think you are to finding the actual height of the object? (Feel free to compare your results to the real answer if you have access to that information.)
- How could you improve your results?

▶ Continued on next page

- How is the process for finding the heights of objects useful in real life?

- Which method would be best to measure the height of an object such as Mount Everest? Explain.

- Do scientific principles lead to good solutions or do good solutions lead to scientific principles? Explain.

③ Linear relationships

Modeling with linear functions can help you understand the interconnectedness of life on Earth and the effects of human activity on this planet. However, it could also help study a wide range of topics, from sports competitions to the future of society.

Identities and relationships

Teams and competition

What does it take to win the 100 m race at the Olympics? How fast are sprinters? Is their speed constant or does it change over the course of the race? Representing this information graphically can help to analyse the contest, and predict future times in the same event.

Other competitions, such as the long jump, rowing, skiing and even archery all have elements that can be graphed, analysed and predicted. The study of linear relationships could become a study of some of the greatest races of all time or help you train in your quest to become the next great athlete.

Fairness and development

Imagining a hopeful future

What do the gray wolf, the American alligator, the peregrine falcon, the snow leopard and the whooping crane all have in common? They have all come back from the brink of extinction. In the past, human activity has endangered many species, but there is hope that we can reverse the trend and make up for our mistakes. Linear relationships can be useful tools in studying these patterns of decline and recovery.

At the same time, there have been some positive trends in our daily lives that seem to indicate a brighter future. The wage gap between men and women is decreasing, medical breakthroughs mean that deaths from cholera and cancer are dramatically decreasing, and rights are increasingly being recognized for all humans, regardless of gender, sexual orientation, race and religion. Analysing and describing these trends with linear relationships could allow you to imagine a fairer future for all inhabitants of the planet.

3 Linear relationships
Impact of human decision-making

Related concepts: Change, Models, Representation

Global context:

In this unit, you will explore how human behaviour affects all aspects of life on our planet. As you extend your study of **globalization and sustainability**, you will see that analyzing data and modeling relationships can help you determine trends. These patterns can then help you to understand the interconnectedness of our planet and how our decisions and actions impact others and the environment.

Statement of Inquiry

Representing patterns of change as relationships can help determine the impact of human decision-making on the environment.

Objectives

- Representing linear relationships in different ways
- Determining the characteristics of a linear relationship (gradient, y-intercept)
- Graphing linear relationships using a variety of methods
- Understanding the relationship between parallel and perpendicular lines
- Applying mathematical strategies to solve problems using a linear model

Inquiry questions

F What is a pattern?
How do you know when something is changing?

C How can you represent changing relationships?
What makes a good representation?

D How does human decision-making affect the environment?
How are we held accountable for our decisions?

You should already know how to:

1 Draw a graph with a restricted domain

Draw the following graphs. Be sure to restrict your diagram to the given values of x.

$y = 2x, 2 \leq x \leq 8$ $\qquad$ $y = x + 3, 4 \leq x \leq 10$

2 Identify the properties of parallelograms and squares

What are the properties of parallelograms? What are the properties of squares?

3 Substitute values into an expression.

If $a = 4$ and $b = -2$, find the value of the following:

$3a + 4b$ $\qquad$ $-a + 5b$ $\qquad$ $-3a - 6b$

4 Solve equations (including square root, fractions, squares)

Solve the following equations.

$6x - 7 = 11$ $\qquad$ $2x^2 = 8$

$\sqrt{x} - 4 = 1$ $\qquad$ $\dfrac{x-5}{3} = \dfrac{1}{2}$

5 Isolate a variable

Isolate y in each of the following.

$6x - 5y = 11$ $\qquad$ $2(y - 5) = 8x$

$y - 4 = \dfrac{1}{2}(x + 3)$

6 Use properties of similar triangles to solve problems

Identify the triangles that are similar. Find the side lengths marked x and y.

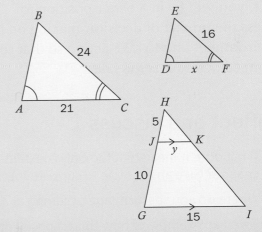

Introducing linear relationships

The Greek philosopher Heraclitus once said that the only constant thing in life is change. What was he talking about? What does it mean to be constant? Can change occur at a constant rate and, if so, does that mean that the future can be predicted?

The type of relationship where the rate of change is constant is called a *linear relationship*. Any change in one variable will always produce a corresponding, predictable change in the other variable. Linear relationships are found all over the world, from the eruptions of a volcano to the build-up of greenhouse gases due to human activity and the foods we consume.

A cow releases 400 liters of methane (a greenhouse gas) a day, mainly by burping. This is the largest amount released by all the livestock animals we raise on farms.

Representing linear relationships both graphically and algebraically will allow you to model real-life situations and discover and reflect on the opportunities and tensions that result from worldwide interconnectedness. By the end of this unit, you will be in a better position to assess the impact of decision-making on humankind and the environment and to take action so that some patterns can be changed before it is too late.

Reflect and discuss 1

- What does it mean that the only constant thing in the world is change? Give an example to support your answer.

- Where do you see a change occurring at a constant rate in your life?

- Given an example of how the actions of others affect you directly. Consider actions by people on a local or even on a national level.

- How are you held accountable for your decisions?

Linear relationships

When a relationship exists between two variables, a change in one is often followed by a change in the other. You have seen in a previous course that the graph of a relationship with a constant rate of change is a straight line. This is called a *linear relationship*. What are the characteristics of linear relationships and what are the different ways that they can be represented?

Representing linear relationships

Linear relationships can be represented in different ways, some of which are explored in the next activity.

Activity 1 – Multiple representations

1 Look at the following representations. Write down those that represent the same relationship. Justify your reasoning.

a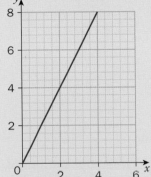

b (0, 1)

(1, 4)

(2, 7)

(3, 10)

c (0, 0)

(1, 2)

(2, 4)

(3, 6)

d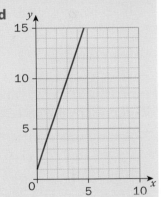

e "Though it had 10% battery life left, Brice's phone was losing $\frac{1}{2}$% every hour."

f "The temperature was 0 °C and is increasing by 2 °C every hour."

g

x	0	2	4	6
y	10	9	8	7

h

x	0	1	2	3	4
y	1	4	7	10	13

i "Starting with one bacterium, the number of bacteria is increasing by 3 every minute."

j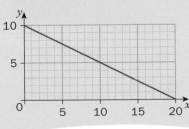

k (0, 10)

(4, 8)

(6, 7)

(8, 6)

▶ Continued on next page

2 Describe the different ways that linear relationships can be represented.

3 Describe how each representation demonstrates the same linear relationship.

4 Why is "linear" an appropriate name for this type of relationship?

Each representation has its advantages and disadvantages and your choice of which one to use will depend on the information you are analyzing and what you are trying to achieve.

Activity 2 – How much water does a leaking faucet waste?

A leaking faucet may not seem like a huge waste of water, but it certainly can add up if you leave it unfixed for a period of time. On average, a leaking faucet drips 15 milliliters of water per minute.

1 Represent this pattern with a table of values. Create a table like the one below, using 1-minute intervals for the x values up to a maximum of 10 minutes.

x (minutes)	0	1	2	3	4	5
y (milliliters)	0	15				

2 Represent the pattern as a set of coordinates.

3 Represent the pattern as a graph for these values of x and y:
$0 \le x \le 10, 0 \le y \le 200$ (Please use a ruler when connecting the points.)

4 Represent the pattern as a verbal description of the relationship between the amount of water wasted (in millilitres) and the time (measured in intervals of 10 minutes).

5 Show that each representation will give the same amount of wasted water after 1 hour.

6 Set up a table showing the same linear relationship but using weeks as the time interval. How much water would be wasted in the course of a year?

7 Most people use about 110 liters of water in a bath. How many baths could be filled using the water from a faucet that has been leaking for one year? Show your working.

Reflect and discuss 2

- What does a linear pattern look like in each representation (verbal description, table of values, coordinates, graph)?
- Name one advantage and one disadvantage of each representation.
- Which representation do you prefer? Explain.

In Activity 2, you were given a verbal description of a constant increase and, from the information given, you were able to generate a table of values, a set of coordinates and a straight-line graph. These are all important representations of a linear relationship. It is essential to not only recognize linear relationships, but to be able to move between these different representations.

Practice 1

1 Match each graph to the appropriate verbal description. Then state the appropriate label (including units) for the axes in each question.

a
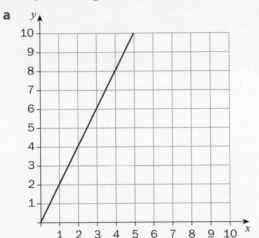

b
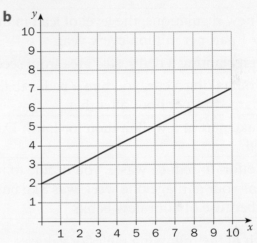

c
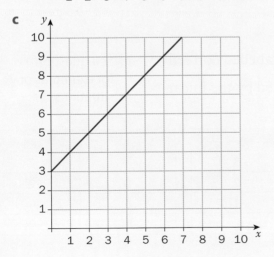

d
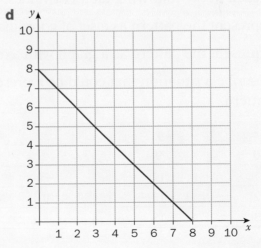

▶ Continued on next page

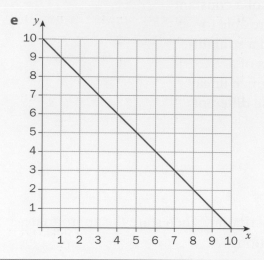

i	After new regulations, the level of lead in the soil, which was 10 parts per million, fell by 1 part per million every year.
ii	When replanting a deforested area from seed, the new plants grow 2 cm each month.
iii	To avoid taking a car, you hire a pedi-cab that has an initial cost of $3 and then you are charged $1 per hour after that.
iv	You take part in a walk-a-thon for Earth Day and are pledged €2 plus €0.50 per km walked.
v	In an effort to reduce waste, you decide to reduce, reuse, recycle and compost. The mass of your garbage one week was 8 kg but it has been decreasing by 1 kg every week because of your new habits.

2 For each of the following tables of values:

a represent the pattern as a set of coordinates

b represent the pattern as a graph with axes labelled correctly

c describe a situation that could be represented by the linear pattern.

> See question 1 for examples of situations.

x	y
0	9
1	11
2	13
3	15

x	y
−6	−3
−3	−2
0	−1
3	0

x	y
−5	7
0	2
6	−4
10	−8

▶ Continued on next page

3 The increase in demand for cars that run on electricity or solar power has begun to reduce our emission of greenhouse gases. Electric cars run on lithium-ion batteries, the cost of which has been dropping so dramatically that they are becoming more affordable to more people. In 2010, the cost of a lithium-ion battery was $1000/kWh. (A kilowatt-hour, or kWh, is a unit of energy. A 1000-watt appliance used for 1 hour uses 1kWh of energy.) The price of these batteries has been decreasing by $125/kWh every year.

a Represent this pattern with a table of values from 2010 until 2017.

b Represent the pattern as a set of coordinates.

c Represent the pattern as a graph. State your domain and justify your boundaries.

> The domain is the set of all x-values.

d Represent the pattern as a verbal description of the linear relationship.

e What do you think will eventually happen to the price of lithium-ion batteries? Will the pattern continue forever? Explain your thinking.

4 Think of a real life situation that can be expressed as a linear relationship.

a Represent this pattern with a table of values.

b Represent the pattern as a set of coordinates

c Represent the pattern as a graph. State your domain and justify your boundaries.

d Represent the pattern as a verbal description of the linear relationship.

e Make a prediction based on your relationship and show that each representation will give the same result for your given domain.

5 A daily decision that most people have to make is what to do with food that they do not eat. Approximately 670 million tonnes of food are wasted each year in high-income countries and 630 million tonnes in low-income countries.

a Set up a table of values to represent the total amount of food wasted over the next 5 years for both high- and low-income countries.

b Between now and the year 2030, how many tonnes of food would be wasted in total in each type of country if this trend continues?

c How does your representation of the linear relationship help you to find the answer to question 5b?

d Is there anything about the question or your answers that surprises you? Explain.

ATL1 **e** What are the obstacles and challenges preventing people from **not** wasting food?

Characteristics of linear relationships

Linear relationships have very specific characteristics which help define them. One of them relates to the steepness of the graph while others are specific points on the graph.

Rate of change

An important property of a line is its *rate of change*. The definition of a linear relationship is that its rate of change is constant.

Constant rate of change (linear relationship)

Time (years)	Ozone (DU)
0	100
1	104
2	108
3	112
4	116

Year	Air quality ($\mu g/m^3$)
1990	120
1995	111
2000	102
2005	93
2010	84

Population (millions)	Area of forest cut (km^2)
10	1100
11	1500
13	2300
15	3100
19	4700
24	6700

NOT constant rate of change (non-linear relationship)

Time (years)	Population (billions)
0	7
1	8
2	10
3	13
4	17

Population (millions)	e-waste (millions of tonnes)
65	5.2
70	6.1
75	6.5
80	7.2
85	7.4

Year	Population (billions)
1804	1
1927	2
1959	3
1974	4
1987	5
1999	6
2011	7

Reflect and discuss 3

- Based on the above examples, explain what it means to have a constant rate of change.

- In which example was the rate of change less obvious? How were you able you tell that the rate was constant?

- How does having a constant rate of change allow you to make predictions? Explain with an example.

> The rate of change is often referred to as the slope or gradient of the line. It is a measure of the steepness of the line.

If linear relationships have a constant rate of change, how do you calculate it? What does it mean?

 # Investigation 1 – Gradient of a line

Incandescent bulbs were the standard light bulb for almost 150 years. However, they are very inefficient and use much more electricity than a light emitting diode (LED) bulb. You can choose to replace your regular 60 W bulb with a 10 W LED bulb, which uses one-sixth of the power and will also last 20 times longer! This will not only save you money, but it will be more beneficial to the planet.

criterion **B**

1 An LED bulb uses 10 joules of energy every second that it is on. Represent the relationship between the number of joules used by the bulb for the first minute that it is on using a table and a graph.

2 Choose any two points on the line and make a triangle showing the horizontal and vertical distances. The slope is the change in the vertical distance divided by the change in the horizontal distance and is defined as $\dfrac{\text{change in } y}{\text{change in } x}$. Determine the slope of the line you drew in step 1.

3 Choose two other points on the line and find the slope. What do you notice? Does that make sense? Explain.

4 Choose a third set of points on the line and verify your conclusion in step 3.

By changing from incandescent bulbs to LED bulbs, Maryam's electricity usage decreased, as shown in the graph below.

5 Choose three pairs of points on the line and calculate the gradient.

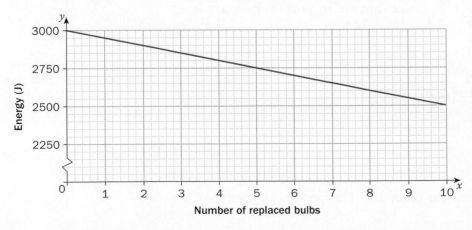

▶ Continued on next page

6 How was the slope of this line different than the slope of the previous one? Does this make sense? Explain.

7 Based on your work, write down a formula to calculate the gradient of a line when you know two points on it. Use variables to represent the coordinates of the points.

8 Verify your rule for a new pair of points on each graph in this investigation.

9 Justify why your formula works.

Reflect and discuss 4

- What do you think is the gradient of a horizontal line? Explain.

- Why is the slope of a line $\dfrac{\text{change in } y}{\text{change in } x}$ rather than $\dfrac{\text{change in } x}{\text{change in } y}$? What difficulties would you encounter if you used the second definition instead?

- The gradient is sometimes said to be $\dfrac{\text{rise}}{\text{run}}$. Does this make sense? Explain.

- Which line is steeper, a line with a gradient of 5 or a line with a gradient of −5? Explain.

Parallel and perpendicular lines

Two or more linear relationships together are referred to as a *system of linear equations*. This will be the focus of Unit 7. However, two specific cases that are important here are lines that are parallel to each other and lines that are perpendicular to each another.

Reflect and discuss 5

Pairs

- In pairs, discuss the meaning of parallel lines and the meaning of perpendicular lines.

- How do you think the gradients of parallel lines compare? Explain.

- How do you think the gradients of perpendicular lines compare? Explain.

Investigation 2 – Parallel and perpendicular lines

criterion **B**

The following diagram shows a parallelogram.

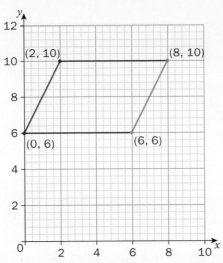

1 What are the properties of a parallelogram?

2 Prove that the diagram is actually a parallelogram by calculating the slopes of its sides.

3 Write a generalization about the gradients of parallel lines.

The following diagram shows a square.

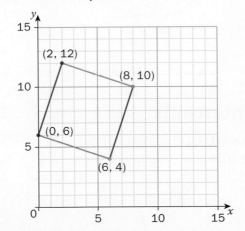

4 What are the properties of a square?

5 You know that the angles in a square are all right angles. Use this to help you determine how the slopes of perpendicular lines compare.

6 Write a generalization about the gradients of perpendicular lines.

7 Verify your generalizations about parallel and perpendicular lines using one more example of each.

8 Justify why your generalizations make sense.

Reflect and discuss 6

Pairs

- Were your initial conjectures about the slopes of parallel and perpendicular lines correct?

- Given that the gradient of a line is a measure of its steepness, explain why what you found out about the gradients of parallel lines makes sense.

- Danika says: "The product of the gradients of two perpendicular lines is always −1." Is she correct? Explain.

Intercepts

Other important characteristics of a linear relationship are its intercepts. These are the points where the line passes through or touches the *x*-axis (*x-intercept*) and the *y*-axis (*y-intercept*). Graphs of linear relationships are not the only types of graphs that have *x*- and *y*-intercepts. Any graph that passes through or touches the *x*- and *y*-axes has these intercepts.

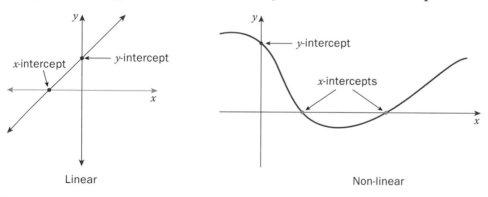

Linear   Non-linear

Investigation 3 – Finding intercepts

1 For each of the graphs below, write down the coordinates of the *x*-intercept and the *y*-intercept. Summarize your results in a table.

a

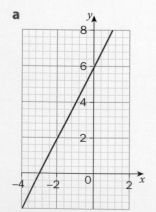

b

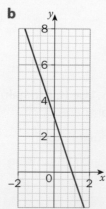

c

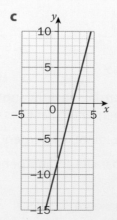

d

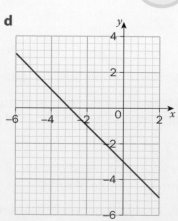

▶ Continued on next page

2 What characteristic do all x-intercepts have in common? What characteristic do all y-intercepts have in common?

3 How can you use your result from step 2 to find the x- and y-intercept in the following equation?

$$2x + 3y = 6$$

4 Show that a relationship defined by $x^2 + y^2 = 4$ has x-intercepts at (0, 2) and (0, −2).

5 A relationship is defined by $x^2 + y^2 = 4$. Find its y-intercepts. Verify your answer by using an online graphing tool like Desmos.

6 Generalize a method for finding the x- and y-intercepts of a relationship given its graph.

7 Generalize a method for finding the x- and y-intercepts of a relationship given its equation.

8 Verify your method by finding the x- and y-intercepts for the relationship defined by $4x − 2y^3 = −16$.

9 Justify why your method works.

In linear relationships, the y-intercept is often named the *initial condition*. For the leaking faucet activity (see page 106), the initial condition was 0, as no water has been wasted when no time has passed. For Maryam's electricity usage (see pages 111–112), when she has replaced zero bulbs with LED bulbs, she is using 3000 J of energy per day.

Activity 3 – Saving water

Installing rainwater tanks near a house is a popular decision in countries like Australia where water shortages can occur. The water from these tanks can be used for watering gardens and even some household applications such as flushing the toilet and washing the laundry if plumbing is installed.

If a small rainwater tank holds 2000 L and is currently full, how many loads of laundry can be done if an average load uses 50 L of water?

x (number of loads)	0					
y (liters remaining in tank)	2000					

▶ Continued on next page

1 Represent this pattern with a table of values.

2 Represent the pattern as a graph: $0 \leq x \leq 50, 0 \leq y \leq 2000$.
Plot the points and then extend the line on your graph using a ruler.

3 Calculate the rate of change and state the initial condition
(y-intercept) in this scenario.

4 Write down the initial condition using coordinates.

5 How many loads can be washed before the tank is empty?
Show your working.

6 Approximate the number of loads of laundry your household does
in a week. How many weeks would the rainwater from a full tank
last in your household?

ATL1   **7** Identify one obstacle and one challenge to installing a rainwater
tank in your home or school.

Practice 2

1 Indicate whether the relationship represented in each table is linear or non-linear.
For those that are linear, find the rate of change.

a

x	y
12	32
14	37
16	42
18	47
20	52

b

x	y
9	78
10	72
12	60
16	36
22	0

c

x	y
2	33
5	36
8	24
11	48
14	60

d

x	y
−2	0
1	9
4	18
8	30
13	45

2 Find the slope of each line $\overleftrightarrow{AB}$.

a $A(-3, -8)$ and $B(12, 10)$

b $A(10, 24)$ and $B(-5, 18)$

c $A(1, -20)$ and $B(\frac{1}{2}, -2)$

d $A(14, 1)$ and $B(-7, 11)$

e $A(-7, 5)$ and $B(3, 5)$

▶ Continued on next page

3 Are the lines shown on the grid below perpendicular? Justify your answer.

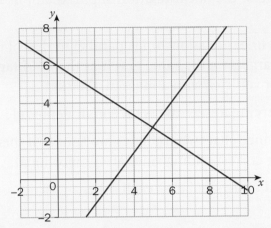

4 You want to draw a square on the Cartesian coordinate plane and you are given the coordinates $A(-5,1)$, $B(-4,4)$, $C(-1,3)$.

  a Draw these three points on a Cartesian grid.

  b Where should the fourth point be placed in order to make the square? Draw the point and label it D.

  c Verify that you have chosen the correct point using two different methods.

5 Find the x- and y-intercepts of the following relationships. (There may be more than one.)

  a $5x - 6y = 30$ **b** $y = -3x + 11$ **c** $2x^2 - 4y = 18$

  d $y = \dfrac{-5}{7}x + 10$ **e** $3x^2 + 2y^2 = 24$ **f** $2\sqrt{x} + 6y = 8$

6 The countries of the European Union made the decision to tackle rising greenhouse gas levels by setting targets for the emission of carbon dioxide (CO_2), one of the most prevalent greenhouse gases. The target CO_2 emission levels for vehicles in 2015 depended on the weight of the vehicle and followed a linear pattern.

2015		2020	
Weight of vehicle (kg)	**CO_2 emissions (g/km)**	**Weight of vehicle (kg)**	**CO_2 emissions (g/km)**
1300	125	1500	100
1800	150	1800	

The targets for 2020 are even more ambitious, though the line is parallel to that of the 2015 targets.

▶ Continued on next page

a Represent the 2015 targets on a graph.

b Determine the rate of change of the linear relationship. Explain its meaning in this context.

c Using the fact that the 2020 linear relationship is parallel to that of the 2015 targets, represent the 2020 targets on the same graph as the 2015 targets to find the emissions target for 2020 for vehicles with a mass of 1800 kg.

ATL1 **d** What are the challenges of trying to set targets on a continent-wide scale?

7 In a previous unit, you determined that two triangles are similar if their angles are exactly the same size and their corresponding sides are proportional.

a Show that triangles *A*, *B* and *C* are similar to each other.

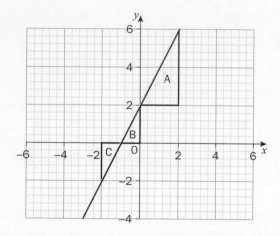

b Explain how these similar triangles can help you to demonstrate that a line has a constant rate of change (that the slope between any two points is always the same).

Algebraic representations of linear relationships

You have seen how you can represent a linear relationship with a graph, a table, coordinates and a verbal description. One of the most important representations is the algebraic representation, which is the focus of this section.

Recognizing linear relationships

What does the equation of a line look like? Is it possible to recognize a linear relationship without graphing it?

 ## Investigation 4 – What is linear?

 criterion **B**

1 Make a table with two columns, one labeled "Linear" and the other labelled "Non-linear". You will need about 15 rows.

Linear	Non-linear

2 Using a graphing display calculator (GDC) or graphing tool like Desmos, draw the graphs of the following equations one at a time. Every time you find one that is a line, write its equation in the "Linear" column. Write the equations of the non-linear graphs in the other column.

a $y = x^2 - 5$ **b** $y = 3x + 9$ **c** $y = \sqrt{2x - 8}$ **d** $y = \dfrac{4}{x}$

e $y = \dfrac{1}{2}x - 3$ **f** $y = 9 - x$ **g** $y = 2x^3 - x + 4$ **h** $y = (x+2)(x-4)$

i $y = -6x$ **j** $y = 2^x + 5$ **k** $y = \sin x$ **l** $y = 1.2x + 24$

m $y = \dfrac{x+3}{x-5}$ **n** $y = \dfrac{2x-6}{3}$ **o** $y = |x-4| + 3$ **p** $y = \dfrac{x}{4} - 2$

q $y = \dfrac{-2}{x^2}$ **r** $y = x$ **s** $y = 3(2x - 12)$ **t** $y = \dfrac{1}{\sqrt{x+2}}$

3 Generalize your results and write a rule for how you can tell if an equation represents a straight line or not.

4 Verify your rule with two more examples.

5 Justify why your rule works.

Reflect and discuss 7

 Pairs

In pairs, discuss whether or not the following equations represent linear relationships. Justify each answer.

- $4x - 2y = 10$
- $2x^2 + 3y^2 = 8$
- $\dfrac{2x}{3} + \dfrac{5y}{2} = -4$
- $y = -4x^3 + 1$

Gradient–intercept form

While all linear relationships have a constant rate of change, they can be represented in a variety of algebraic forms, such as *gradient-intercept form*.

Investigation 5 – What does this number do?

Online applications, such as the ones on the Math is Fun website or on desmos.com, will allow you to perform this investigation using sliders to alter the parameters m and c. You can use these online tools, a graphic display calculator (GDC) or graphing software.

1 Using sliders or a graphing tool, change the value of m in $y = mx$. Note the changes in the graph.

2 Using sliders or a graphing tool, select a value for m and then change the value of c in $y = mx + c$. Note the changes in the graph.

3 Using sliders or a graphing tool, change the values of both m and c in $y = mx + c$. Note the changes in the graph.

4 Generalize your results on the effects of m and c and summarize them in a table. Be sure to include the following details.

- What do m and c refer to in terms of the components of a linear relationship?

- Which parameter, m or c, affects the direction of the graph of $y = mx + c$ and in what way(s)?

- Which parameter, m or c, affects the steepness of the graph of $y = mx + c$ and in what way(s)?

- What is the effect of changing c on the graph of a linear relationship?

5 Verify your results for two more linear relationships: $y = -\dfrac{1}{2}x + 4$ and $y = 3x - 5$.

6 Justify why your results make sense.

Representing lines in gradient–intercept form, $y = mx + c$, has advantages when it comes to graphing these relationships. The slope and the y-intercept can become tools that allow you to draw the graph without having to use a table of values or coordinates.

Activity 4 – Graphing a line

Use the following equations in this activity.

Equation	$y = 3x$	$y = 2x - 1$	$y = -x + 4$	$y = -2x - 4$	$y = \frac{1}{2}x$	$y = -4x + 6$

1 Without using technology, make a prediction about where each of these graphs crosses the y-axis, the direction of the graph and its steepness (how many units vertically and how many units horizontally).

2 Using the information from step 1, graph each of these equations on a set of axis ranging from −10 to +10.

3 Verify your graphs using technology.

4 Summarize the steps you can follow in order to graph a linear relationship.

Example 1

Q For the line given by $y = 4x - 2$

    a State its slope and its y-intercept.

    b Graph the linear relationship.

A   **a** The slope is given by m, so this is 4.

       The y-intercept is (0, −2).

> The equation of the line is in gradient-intercept form, $y = mx + c$.

    b

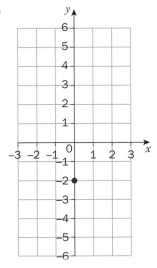

> When the equation is in slope–intercept form, you start with the value of the y-intercept and plot that point on the y-axis In this case, plot the point (0, −2).

▶ Continued on next page

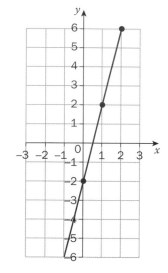

Gradient $= 4$

$$= \frac{4}{1}$$

slope $= \dfrac{\text{rise}}{\text{run}}$ so you convert the slope (m) into a fraction.
Any whole number is represented with 1 as the denominator. If the slope is negative, put the negative sign with the numerator.

From the y-intercept you go up for the rise (or down for a negative number) and to the right for the run. In this case, you will go four units up and one to the right.

Continue this process to keep plotting points. You can also plot a point in the opposite direction from the y-intercept (in this case by going four units down and one unit to the left).
Then use a ruler to draw the straight line connecting all the plotted points.

Practice 3

1 Graph the following linear relationships using the gradient and y-intercept.

a $y = x$

b $y = -3x + 7$

c $y = 5x - 3$

d $y = -\dfrac{1}{2}x - 4$

e $y = \dfrac{2}{3}x + 1$

f $y = -4x$

g $y = 2x + \dfrac{1}{2}$

h $y = 3 - \dfrac{2}{5}x$

i $y = \dfrac{-x}{4}$

2 Which of the following lines are parallel? Which are perpendicular? Justify your answers.

a $y = 5x - 4$

b $y = -5x - 4$

c $y = \dfrac{4}{3}x + 10$

d $y = \dfrac{1}{5}x$

e $y = \dfrac{3}{4}x + 1$

f $y = -\dfrac{1}{5}x + 3$

g $y = -5x + 11$

h $y = -\dfrac{3}{4}x - 7$

i $y = \dfrac{-4x}{3} + 10$

▶ Continued on next page

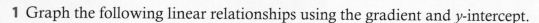

3 As a way of trying to limit carbon emissions in the transportation sector, train companies are changing to eco-friendly hybrid locomotives which reduce emissions by 70%. A hybrid diesel electric locomotive has an average speed of 80 km per hour.

a Copy and complete this table of values.

Time travelled in hours (x)	0	1	2	3	4
Distance travelled in kilometres (km)					

b Calculate the rate of change and the initial condition of this linear relationship.

c Represent the table of values as a graph.

d Use your graph to find how far the locomotive travels in 3.5 hours.

e Use the rate of change and initial condition to represent the linear relationship as an equation. Use your equation to verify your answer in step d.

f Explain in your own words what the rate of change and initial condition mean in the context of this situation.

4 Concrete is the world's most widely used material (apart from water) due to its versatility in construction. The drawback is that, for every tonne of concrete produced, 0.82 tonne of CO_2 is released. Increased CO_2 levels have been linked to global warming and climate change. Concrete accounts for almost 10% of the world's CO_2 emissions.

a Represent the relationship between tonnes of concrete produced and tonnes of CO_2 released for up to 100 tonnes of concrete.

b Find the rate of change and the initial condition.

c Write down the equation of this linear relationship.

d Use your equation to calculate how much concrete has to be made to release 205 000 tonnes of CO_2.

e More than 4 billion tonnes of concrete are produced each year. How much CO_2 does this release annually?

5 The technology exists to cut CO_2 emissions by up to 90% by using different concrete recipes. One developer claims it has created a concrete that can **absorb** CO_2 from the air. When water is added, every tonne of concrete absorbs 0.6 tonnes of CO_2.

a Write an equation to represent this scenario.

b If the whole world switched to this type of concrete, how much CO_2 would be absorbed by the 4 billion tonnes of concrete used each year?

c Some of these innovative concrete varieties have been available for years but have not been popular, and some of the technology has been abandoned because of lack of funds. Why do you think this is? What action needs to take place?

ATL1

d What obstacles and challenges stand in the way of this new technology being used for all concrete?

▶ Continued on next page

6 The bottled water industry is currently worth over $200 billion annually. The average cost of a bottle of water in the United States is 50 cents.

a Determine the equation that represents the cost (y) of buying x bottles of water. What does the gradient represent in this equation?

b What is the total cost of buying one bottle of water a day for a year (not a leap year)?

c A high-quality, refillable water bottle costs approximately $40. How many bottles of water would you have to drink in order for this option to be more cost effective?

d What other factors, apart from cost, might inflence someone's decision whether to carry their own water bottle or to buy bottles of water?

7 The following table contains the capital cost (cost to build the plant before creating any power) and operating cost for a coal power plant and a hydroelectric power plant.

	Capital cost	Operating cost (per kWh)
Coal	$80 000 000	$0.20
Hydroelectricity	$250 000 000	$0.01

a Determine the equation that represents the cost for each type of plant. What do the gradient and y-intercept represent in these equations?

b What is the cost of producing 5000 kWh of electricity at each plant?

c What is the cost of producing 10 000 kWh of electricity at each plant?

> The total cost of electricity is the cost to build the facility plus the cost to produce electricity. You can find the cost of producing electricity by multiplying the operating cost per kWh by the number of kWh the facility generates.

d Coal plants need fuel, whereas hydroelectric plants do not. What will happen to the linear cost functions of both plants (gradient and y-intercept) if the price of fuel increases?

e What other factors, apart from cost, might influence a government's decision-making when considering what type of power plant to build in its country?

Other algebraic representations of linear equations

The equation of a line can be given in a variety of forms, one of which is gradient–intercept form. The equation of a line with slope m, y-intercept c and passing through point (x_1, y_1) can be represented in the following ways:

Slope–intercept: $y = mx + c$

Point–slope: $y - y_1 = m(x - x_1)$

Standard form: $Ax + By = C$, where A, B and C are integers.

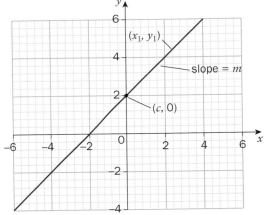

Activity 5 – Moving between different forms

1 Copy and complete this table. Given a point and the gradient of a linear relationship, represent its equation in the three different forms.

Point and gradient	Point–slope form $y - y_1 = m(x - x_1)$	Gradient-intercept form $y = mx + c$	Standard form $Ax + By = C$
$(1, 3)$ and $m = 2$	$y - 3 = 2(x - 1)$	$y - 3 = 2x - 2$ $y = 2x + 1$	$y = 2x + 1$ $-2x + y = 1$
$(-4, -7)$ and $m = \dfrac{1}{2}$			
$(2, -1)$ and $m = -4$			
$(0, 5)$ and $m = -1$			
$(-6, 4)$ and $m = \dfrac{2}{3}$			
$(-9, 2)$ and $m = -7$			

2 Which form do you prefer? Explain.

3 What are some of the advantages and disadvantages of each form?

The form of the equation of a linear relationship may determine the easiest way to graph it since there are also several options for doing this.

Example 2

Q Graph the following line.

$2y + 18 = -3x$

> If you are using graphing software or a GDC, the equation often needs to be in slope–intercept form.

A $2y + 18 = -3x$

$2y = -3x - 18$

$y = \dfrac{-3}{2}x - 9$

> Rearrange the equation to isolate the y.

$m = \dfrac{-3}{2}$ and $c = -9$

> From this, extract m and c.

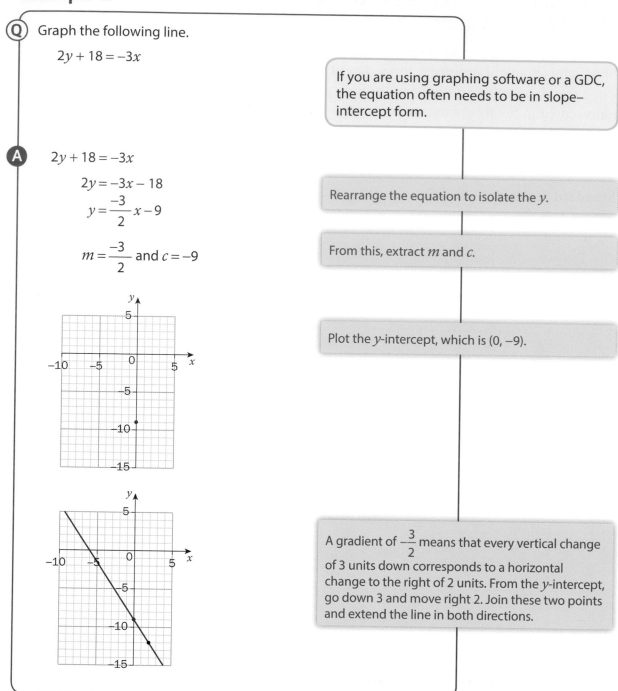

> Plot the y-intercept, which is $(0, -9)$.

> A gradient of $-\dfrac{3}{2}$ means that every vertical change of 3 units down corresponds to a horizontal change to the right of 2 units. From the y-intercept, go down 3 and move right 2. Join these two points and extend the line in both directions.

Are there other ways to draw a line, given the same information?

Example 3

Q Graph the following line.

$$2y + 18 = -3x$$

> Two points define a line. You may find any two points, but the easiest ones tend to be the x- and y-intercepts.

A $2(0) + 18 = -3x$

> To find the x-intercept, substitute $y = 0$.

$$2(0) + 18 = -3x$$
$$18 = -3x$$
$$-6 = x$$
$$(-6, 0)$$

> Solve for x.

> Write down the ordered pair.

$$2y + 18 = -3(0)$$

> To find the y-intercept, substitute $x = 0$.

$$2y + 18 = 0$$
$$2y = -18$$
$$y = -9$$
$$(0, -9)$$

> Solve for y.

> Write down the ordered pair.

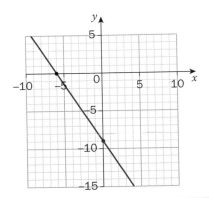

> Plot the two points and draw a straight line to connect them.

Reflect and discuss 8

- Is there such a thing as a **best** method of graphing a particular linear relationship? If so, can you tell from the equation what the best method might be? Explain.

- In which situation(s) is each of the three forms of linear equation most useful? Explain.

- Which method for graphing a line do you prefer? Explain.

Practice 4

1 Represent each of the following linear relationships in the form $y = mx + c$.

a $2y = -4x + 8$

b $5x - 10y = 20$

c $0 = 2x - 3y + 9$

d $\frac{1}{2}y - 3x = 5$

e $\frac{2}{3}y = 2x + 6$

f $4x = 8 - 2y$

g $\frac{3}{2}x - \frac{1}{3}y = 0$

h $\frac{x}{4} + \frac{y}{2} - 1 = 0$

i $\frac{2}{5}y - \frac{1}{10}x = -2$

2 Represent the lines in question 1 on a graph using the gradient–intercept form.

3 Find the x- and y-intercepts for each of the equations in question 1. Validate your graphs by checking to see whether these intercepts are on the lines.

4 Which of the following lines are parallel? Which are perpendicular? Justify your answers.

a $x - 7y = 10$

b $2x - y = 8$

c $3x + 6y = -2$

d $6x + 4y = -5$

e $y = \frac{1}{7}x - 5$

f $8 - 7x = -y$

g $28y = 4x - 12$

h $8x + 4y = -3$

i $y = -2x + 9$

j $7x = 18 - y$

k $4x + 2y = 1$

l $y = \frac{2}{3}x$

5 Graph the following lines by plotting the x- and y-intercepts.

a $3y + 4x = 24$

b $5x - 10y - 30 = 0$

c $0 = 2x - 4y + 12$

d $\frac{y - 3x}{2} = 6$

e $\frac{1}{4}y = 2x + 4$

f $4y = \frac{1}{2}x - 12$

6 Graph the following lines using a method of your choosing.

a $y = 3x - 12$

b $3y = 5x + 15$

c $0 = -3x + 14 - 7y$

d $y = -x + 4$

e $4y - 8x = 12$

f $\frac{y - 2}{3} = -2x$

g $2y = 2x - 10$

h $-2 = 3x - \frac{1}{2}y$

i $\frac{1}{3}x - \frac{2}{5}y = 0$

7 Using asphalt to create roads and parking lots has created "urban heat islands" because the asphalt absorbs the Sun's rays and radiates the heat back into the atmosphere. Temperatures in these zones are much hotter than in suburban and rural areas, in some cases causing severe health issues. To combat this, the city of Los Angeles elected to paint the asphalt with a special material that reflects the Sun's rays.

a Crews can paint $4\,m^2$ each minute. Represent this relationship using a table, a graph and an equation written in all three forms. Show your working.

b What do the gradient and y-intercept mean in the context of this situation?

▶ Continued on next page

c Because of this new technology, temperatures cool off much faster at night than without the new material. Suppose the temperature at 6 pm is 28°C and it drops by 0.5°C every hour. Represent this relationship using a table, a graph and an equation written in all three forms. Show your working.

d What do the gradient and y-intercept of this new equation mean in the context of this situation?

Activity 6 – Turn it "OFF"

The World Wide Fund for Nature issues tips on how to lessen your impact on the environment.

One of these tips is to completely turn off equipment when not in use and to unplug chargers, even if no device is attached. Approximately one-quarter of all residential energy consumption in industrialized countries is used by electrical devices in "idle" power mode. These devices and chargers use what is called "standby power" or "phantom power".

1 Find out the cost per kilowatt hour that your electricity provider charges.

2 Look at your current electricity bill and assume that 25% is wasted on devices in "idle" mode.

3 Create a linear equation (or *linear model*) to represent the cost of phantom power in your home. What is your x variable? What is your y variable?

4 Describe the rate of change and the initial condition in the context of this question.

5 Represent your model as a graph.

6 Use the model to predict how much energy and money you could save in a year if you followed the given tip. Show all your calculations.

ATL2

7 Go through your house and make a list of all devices and cords that draw phantom power. Research ideas from a variety of reputable sources and create an action plan for easy ways to save energy in your home.

ATL1

8 What are the challenges involved in making the choice to turn off or unplug your computer/phone charger/television when you are not using them?

Determining the equation of a line

In order to determine the equation of the line, you simply need a point on the line (x, y) and the gradient of the line (m).

Example 4

Q Find the equation, in slope–intercept form, of a line that passes through (2, 7) and has a slope of 5.

A

$y = mx + c$

$7 = (5)(2) + c$

> Start with the equation and substitute in the values you have been given.

$7 = 10 + c$

$-3 = c$

> Solve for the y-intercept.

$y = 5x - 3$

> Write down the equation in the required form.

Example 5

Q Find the equation, in standard form, of a line that passes through (9, −2) and (−15, 6).

A

$m = \dfrac{y_2 - y_1}{x_2 - x_1}$

> Calculate the slope by using the formula.

$m = \dfrac{6 - (-2)}{-15 - 9}$

$m = \dfrac{8}{-24}$

$m = -\dfrac{1}{3}$

$y = mx + c$

> Substitute the slope and the coordinates of one point into the equation.

$-2 = \left(-\dfrac{1}{3}\right)(9) + c$

$-2 = -3 + c$

$1 = c$

$y = -\dfrac{1}{3}x + 1$

> Write down the equation in gradient–intercept form.

$3y = -x + 3$

$x + 3y = 3$

> Multiply both sides of the equation by 3 so that all coefficients are integers. Then write the equation in the required form.

Example 6

Q Find the equation of the line through $(6, -8)$ that is perpendicular to the line $y = 3x + 5$. Write your answer in slope–intercept form.

> The ∴ symbol means "therefore".

A

$$m = 3$$

$$\therefore m_{\text{perp}} = -\frac{1}{3}$$

> Determine the slope of the linear relationship. If it is parallel to the given line, use the same slope. If it is perpendicular, use the negative reciprocal.

$$y = mx + c$$

$$-8 = \left(-\frac{1}{3}\right)(6) + c$$

$$-8 = -2 + c$$

$$-6 = c$$

> Substitute the slope and coordinates of the point into the equation.

$$y = -\frac{1}{3}x - 6$$

> Write down the equation in the required form.

Practice 5

1 Represent the following linear relationships algebraically. Write your answers in all three forms (gradient–intercept, point–slope and standard form).

a

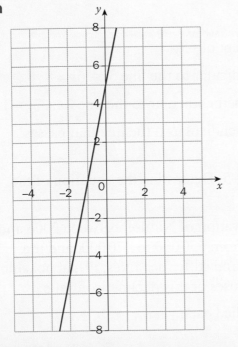

b

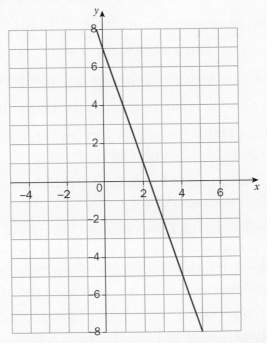

► Continued on next page

c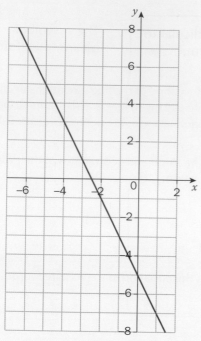

d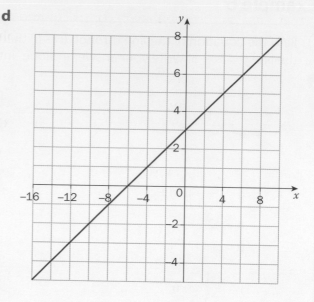

2 Find the equation of the following lines. Write your answers in gradient–intercept form without using decimals.

a The line with a slope of −4 and a y-intercept of 5

b The line with a slope of $\frac{1}{3}$ and passing through (2, 3)

c The line with a slope of −2 and an x-intercept of 5

d The line passing through (−6, −2) and (4, 3)

e The line with a y-intercept of 4 and an x-intercept of −3

f The line with an x-intercept of 7 that is perpendicular to the line $3x - 6y = 8$

g The line that passes through the origin that is perpendicular to the line $2x + 7y = 12$

h The line that passes through (−1, 1) and is perpendicular to the line that passes through (4, −2) and (−3, 1).

3 The line through (2, −4) and (x, 1) has a slope of −5. Determine the equation of the line and the value of x.

4 According to the International Union for Conservation of Nature, between 2006 and 2013 the African elephant population decreased from 550 000 to 470 000, and the number continues to decrease. Despite warnings and stiffer penalties, illegal poaching and selling elephant ivory, which are the main causes of this decline, continue.

a What is the average number of elephants that died each year between 2006 and 2013?

▶ Continued on next page

b Assuming the population continues to decrease at this average rate every year, represent the African elephant population since 2013 with a linear model. Use your equation to predict the African elephant population this year.

> Set 2006 to be year 0 and determine the equation.

c Assuming the population continues to decrease at the same rate, when will the African elephant become extinct?

5 As the Earth heats up, sea levels rise because warmer water takes up more room than colder water, a process known as thermal expansion. Melting glaciers compound the problem by dumping more fresh water into the oceans. In 1993, NASA started measuring sea levels. Since then, the sea level has risen an average of 3.2 millimetres per year.

a Set up a linear equation to model this phenomenon. According to this data, how much has the sea level risen since NASA started recording the data?

b Worldwide, approximately 100 million people live within three feet (94 cm) of sea level. If the current trend continues, when will all of these people's homes be flooded?

 Search for the video entitled "How Earth Would Look If All The Ice Melted". You can see what could happen to major cities all over the world if the polar ice caps and permafrost on Earth melted.

6 This graph represents global carbon dioxide levels in the atmosphere since 1980.

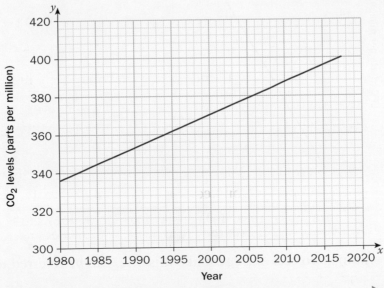

▶ Continued on next page

a According to the graph, what was the CO_2 level in 1990? What was it in 2017?

b According to the graph, what is the change per year of CO_2 levels in the atmosphere?

c Use the graph to determine the equation of the line, using 1980 as your initial condition (y-intercept).

d Verify your equation using different points on the graph. Discuss any discrepancies you may have.

e Researchers have discovered that Antarctica's ice will be significantly more vulnerable to melting once CO_2 levels exceed 600 parts per million. When will this occur if the current trend continues?

ATL2

f Research your country's carbon dioxide emissions per person. How does your country compare with the rest of the world? Why do you think this is?

> Be sure to check a variety of reputable sources. Reference these in your answer and state how you know the source is reliable.

7 Chlorofluorocarbons (CFCs) were created in 1928 and were used in products such as aerosol propellants, cleaning solvents and refrigerants. It has since been proved that CFCs contribute to the destruction of the ozone layer. They continued to accumulate in our atmosphere until a global decision was made to address this potentially catastrophic issue.

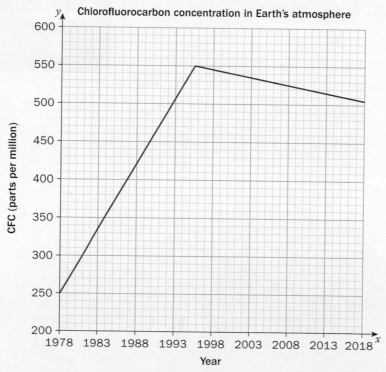

Chlorofluorocarbon concentration in Earth's atmosphere

A graph that contains two sets of trends, like this one, can be referred to as a piecewise relationship.

▶ Continued on next page

134   3 Linear relationships

a Using a variety of sources, research the global effort to ban CFCs that officially started in the late 1980s. Do the sources say it was successful? Explain how you know you can trust the points of view presented.

b Use the graph to determine whether the global agreement to ban CFCs was successful. Explain your answer.

c According to the graph, what was the rate of change in CFC concentrations in the atmosphere leading up to the global ban? Verify your answer using different points on the graph. Discuss any discrepancies you may have.

d According to the graph, what was the rate of change in CFC concentrations in the atmosphere after the global effort started? Verify your answer using different points on the graph. Discuss any discrepancies you may have.

e Although much has been accomplished, CFCs decompose very slowly, so even when all CFC production has stopped (which is not yet the case), some concentration of CFCs will remain in the atmosphere for over 100 years. According to the decreasing trend, in what year will there be zero CFCs left in the atmosphere?

f Write down two facts that you have learned from this case study and its graph.

Formative assessment

criteria
C, D

Reducing the emission of greenhouse gases has been a major focus of governments across the globe since the mid-1990s. In 1997, most countries signed the Kyoto Protocol, agreeing to reduce greenhouse gases and their effect on global temperatures.

The following data represent the CO_2-equivalent concentration of all greenhouse gases in the atmosphere in parts per million over a span of 20 years.

1995	427
2000	441
2005	455
2010	469
2015	483

a Plot these points on a graph, making sure you start your x-axis at 1990 (as that was the benchmark year agreed in the Kyoto Protocol). Plot the years from 1990 to 2030.

b According to your linear model, what was the concentration of the mix of greenhouse gases in 1990?

▶ Continued on next page

c According to your linear model, what is the change per year in the concentration of greenhouse gases in the atmosphere?

d Use your graph to determine the equation of the line, using 1990 as the initial condition. Write the equation in the three different forms.

e Verify your answer to step **d** using algebra.

f According to your model, what is the concentration of greenhouse gases this year?

g If the Kyoto Protocol threshold is 500 ppm for the concentration of greenhouse gases, in what year will the concentration reach that amount if current trends continue?

h Draw this threshold on your graph as a dotted horizontal line. Continue your line to show the point at which it intersects the threshold. Verify that your graphical results are the same as your algebraic results.

Reflect and discuss 9

- What is the Kyoto Protocol, when was it adopted and when was it enforced?

- How many countries agreed to the Kyoto Protocol?

- Given this agreement, would you expect the greenhouse gas emissions data that you graphed in the Formative assessment to continue to increase in the short term? In the long term? At what rate?

ATL1
- What are some obstacles to reaching and enforcing an agreement like the Kyoto Protocol?

- Whose responsibility is reducing greenhouse gases? Who should be held accountable – industry, government or individuals? Explain.

Vertical and horizontal lines

So far, the lines you have studied have all been diagonal or sloping. What happens when the line is vertical or horizontal? Can its steepness be measured? How can you represent these lines with equations?

Investigation 6 – Equations of vertical and horizontal lines

criterion B

1 Graph the following lines by first creating a table of values for each line. Note that $y = 2$ means that the value of y is 2 for any value of x. Graph all three lines on one set of axes.

$y = 2$	
x	y

$y = 3$	
x	y

$y = -5$	
x	y

2 How are the lines related? What type of lines are they?

3 Pick any two points on each line. Calculate the slope of each line.

4 How are the slopes related? How does the slope relate to the steepness of the graph?

5 Repeat step 1 for the following lines. Note that $x = 2$ means that the value of x is 2 for any value of y.

$x = 2$	
x	y

$x = 3$	
x	y

$x = -5$	
x	y

6 How are these lines related? What type of lines are they?

7 Pick any two points on each line. Calculate the slope of each line.

8 How are the slopes related? How does the slope relate to the steepness of the graph?

9 Write a generalization about lines that are of the form $x = a$, where a is a number.

10 Write a generalization about lines that are of the form $y = b$, where b is a number.

11 Verify your generalizations for one more case of each type.

12 Justify why your generalizations are true.

Practice 6

1 Find the equations of the lines shown on the set of axes below.

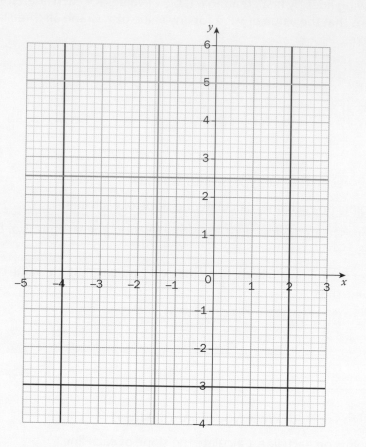

2 Find the equation of the line through (6, 1) that is parallel to the line $y = 34$.

3 Find the equation of the line through (0, 1) that is perpendicular to the line $x = -2$.

4 Find the equation of the line through (5, 5) and (5, −2).

5 Find the equation of the line through (−3, −2) and (1, −2).

6 Find the equation of the line through (4, 2) that is parallel to the line $x = -8$.

7 Find the equation of the line through (−7, 8) that is perpendicular to the line $y = -4$.

Unit summary

Linear relationships can be represented in a variety of ways:

- graph
- table
- coordinates
- words
- algebraic equation.

Some of the characteristics of a linear relationship are its gradient (or slope), its x-intercept and its y-intercept.

The gradient or slope of a line is a measure of its steepness. It is defined as the ratio of its vertical change (rise) to its horizontal change (run). The formula for the gradient of the line joining the point (x_1, y_1) and (x_2, y_2) is

$$m = \frac{y_2 - y_1}{x_2 - x_1}$$

A **positive gradient** slopes upwards from left to right.
A **negative gradient** slopes downwards from left to right.

If two lines are parallel, then their gradients are the same.
If two lines are perpendicular, then their gradients are negative reciprocals of each other (for example, $\frac{2}{5}$ and $-\frac{5}{2}$). Another way to state this relationship is that the product of the slopes of perpendicular lines equals -1. For example $\frac{2}{5}x - \frac{5}{2} = -1$.

The x-intercept is the point where a graph crosses the x-axis. Given the equation of the graph, you can find the x-intercept by substituting $y = 0$ and solving for x.

The y-intercept is the point where a graph crosses the y-axis. This is also referred to as the initial condition in the context of application questions. Given the equation of the graph, you can find the y-intercept by substituting $x = 0$ and solving for y.

To graph a line:

1 Plot the y-intercept on the coordinate axes.
2 Starting from the y-intercept, use the gradient to determine the coordinates of another point on the coordinate axes. Continue this process for a few more points.
3 Draw a line between the y-intercept and the other points.

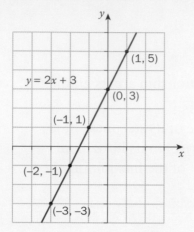

A linear relationship can be represented in one of three forms:

Gradient–intercept or (slope–intercept) form: $y = mx + c$

Point–gradient or (point–slope) form: $y - y_1 = m(x - x_1)$

Standard form: $Ax + By = C$, where A, B and C are integers.

Given the graph of a line, you can determine the equation of the line in two different ways:

- Select two points on the graph, find the gradient and then use the gradient and a point to write the equation in point gradient–form.
- Use the gradient and where it cuts the y-axis to find the y-intercept, and then write the equation in gradient–intercept form.

Horizontal lines have a gradient of zero. The equation of a horizontal line is always of the form $y = a$.

Vertical lines have a gradient that is undefined. The equation of a vertical line is always of the form $x = a$.

Unit review

criterion A

Key to Unit review question levels:

| Level 1–2 | Level 3–4 | Level 5–6 | Level 7–8 |

1. **State** the slope and the y-intercept of each of the following lines.

   a $y = 6x - 12$

   b $y = -\dfrac{2}{7}x + 23$

2. **Determine** the x- and y-intercepts of the following lines.

   a

   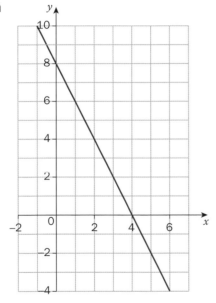

   b $5x - 4y = 20$

   c $7x + 3y = 42$

3. **Plot** the graph of each of the following lines using a method of your choosing.

   a $y = -2x + 8$

   b $y = x - 7$

④ Copy and complete the table for lines A, B, C and D.

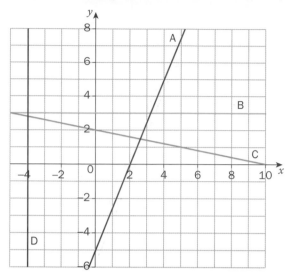

Line	Slope	y-intercept	x-intercept	Equation
A				
B				
C				
D				

⑤ Represent each of the following equations using gradient–intercept form.

a $3y = \dfrac{1}{2}x - 12$ **b** $5x = 10y - 20$ **c** $x = 8y$

d $\dfrac{1}{2}x = 3y - 2$ **e** $\dfrac{2}{3}y - 8x = -6$

⑥ Find the gradient of the line L that passes through each pair of points.

a $A(1, 5)$ and $B(1, -2)$ **b** $A(-1, 6)$ and $B(2, 9)$

c $A(0, 4)$ and $B(2, 8)$ **d** $A(3, 2)$ and $B(5, 4)$

⑦ A tank has a slow leak in it. The water level starts at 100 cm and falls 0.5 cm a day.

a Is this a constant increase or constant decrease situation?

b **Write down** an equation showing the relationship between day, d, and water level, L.

c After how many days will the tank be empty?

8 Graph each of the following lines using a method of your choosing.

a $6y + 3x = 12$

b $10y - x = -50$

c $2y + \dfrac{1}{3}x = 10$

d $2y + 3 = \dfrac{x}{4}$

e $0 = 24 + 8x + 3y$

f $4x = \dfrac{10 + 2y}{3}$

9 Match each of the following equations to its graph.

a $3x - 5y = 15$

b $y = \dfrac{1}{2}x$

c $\dfrac{4}{3}x + \dfrac{1}{2}y + 4 = 0$

i
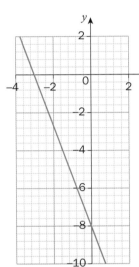

ii
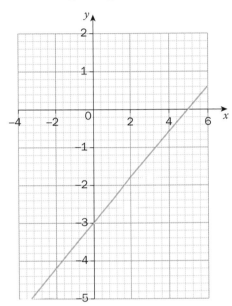

iii
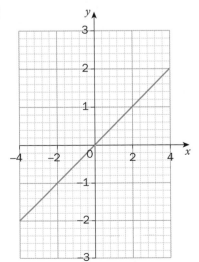

10 The line through (2, −4) and (c, 1) has a slope of −5.
Determine the value of c.

11 Write the equation of the line parallel to $3x + y − 4 = 0$ that
passes through the point (2, −5).

12 Find the equation of a line that goes through (−3, 9) and (9, 1).

13 Write the equation of the line perpendicular to $3x + y − 4 = 0$
that passes through the point (2, −5).

14 Write the equation of the line parallel to $x = −3$ passes
through (6, −7).

15 Determine the equation of the line perpendicular to $y = 4$ that
passes through (−1, 6).

16 In 2016, there were a total of 36.63 million people
already living with HIV, and there were 1.8 million new
HIV infections in that year. The rate of new infections is
decreasing by approximately 300 000 people per year due to
advancements in drug technology and the support of global
agencies and local governments.

  a Write a linear equation that models the
number of people in millions who have been
newly infected with HIV since 2016.

  b If the rate of infections continues to follow
this linear trend, in what year will there be no
new infections?

  c Is this a realistic model to use for such an
epidemic? Why or why not?

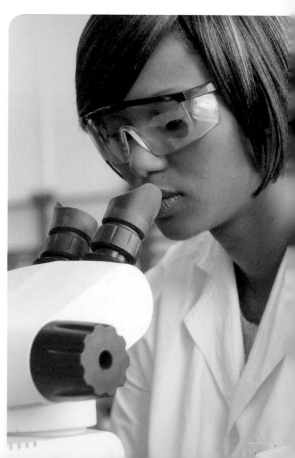

17 According to the National Academy of Sciences of the United States of America, there were 7100 cheetahs left in the wild in 2016. The number had decreased significantly from an estimated 14 000 in 1975, when the last comprehensive count was done in Africa. The decision to convert wilderness areas to agricultural or livestock farms has caused the loss of habitat that is the main cause of this decline.

   a What is the average decrease in the number of cheetahs each year?

   b Assuming this is still the average rate of decline, what is the cheetah population this year?

   c Assuming this average rate of decline continues, when will the cheetah become extinct?

> Set 1975 to be year 0 and determine the equation.

18 Determine the equation of the line perpendicular to the line $4x + 2y - 7 = 0$ that has the same x-intercept as the line $2x + 3y - 12 = 0$.

19 Find the value of k if the lines $3x - 2y - 5 = 0$ and $kx - 6y + 1 = 0$ are:

   a parallel

   b perpendicular.

20 Triangle ABC has vertices $A(3, -1)$, $B(-3, -5)$ and $C(-1, 5)$. Determine whether the triangle is a right triangle. Fully justify your answer.

21 In 1991, 18.6% of the world population was undernourished. In 2015, 10.9% of the population was undernourished.

   a What was the average percentage change per year?

   b Assuming this rate continues, what is the percentage of the population that is undernourished this year?

   c Determine an equation to represent this scenario.

   d Assuming this rate continues, when will there be no undernourished people in the world? Do you think this is actually possible?

   e What decisions at local, country and world level could be driving this change?

22 The decision of which car to buy is no longer just about color and style; it may also involve choosing one that is environmentally friendly. A cost comparison of cars is now a necessity when shopping around for a new car. In the United States, the Ford Focus sells for $19 000 and costs 8 cents per mile to run. The zero-emission electric Focus sells for $21 600 and costs only 3 cents per mile to run.

a Determine the equation that represents the relationship between the total cost of each car (including both buying and running costs) and the miles traveled. What do the gradient and y-intercept represent in these equations?

b What is the total cost of each car at 30 000 miles? If you were planning to sell the car once it reached 30 000 miles, which car would be the cheapest option?

c What is the total cost of each car at 60 000 miles? If you were planning to sell the car once it reached 60 000 miles, which car would be the cheapest option?

d Which characteristic of these linear relationships (gradient or y-intercept) would be affected if

  i the price of gasoline in the US increases?

  ii the US government offers a larger rebate (a partial refund) on the purchase of new electric cars?

e When is the total cost of each car exactly the same? Write a general statement relating the mileage of the cars to when each model is the cheapest, using your original equations.

f What other factors apart from cost might go into your decision of which car to buy?

23 On average, 3.85 kilograms of feed are needed for every kilogram of meat we consume (this is called the *feed conversion ratio*).

a Represent this information in a table showing the relationship between kg of feed and kg of animal meat. Your table must have at least 6 rows of data.

b **Sketch** this linear model, with kg of feed on the x-axis and kg of animal meat on the y-axis. Even if your data doesn't go that high, make sure your x-axis goes to 130 kg.

c Research the most recent meat consumption per capita (per person) data for your country. **Draw** a horizontal line on your graph to represent this.

d How many kilograms of feed are needed to sustain the livestock raised for one person's meat consumption in your country?

e Research the population of your country. What is the overall weight of the feed needed to meet your country's annual meat demands?

f If corn crops yield approximately 4445 kg of grain per acre annually and 60% of that is fed to livestock, how many acres need to be planted with corn in order to feed the livestock eaten in your country?

g To put this area in perspective, research the area of the city or town in which you live. How does it compare with the area of farmland needed to feed the livestock?

Did you know...?

Edible insects are a sustainable, ecologically viable meat source and could actually be a feasible replacement for some of the meat we eat. Crickets, for example, have a feed conversion ratio of 1 to 1 so they are more efficient at producing meat than any of the typical animals we eat. The average edible insect is also approximately 50% protein, compared with the lean cuts of the big four meats which range from 28% to 34%.

criteria
C, D

Challenges of feeding a growing planet...

By 2030, it is estimated that we will need to feed a world population of 8.5 billion people. How can we do that in a way that is environmentally sustainable?

You have been asked to be a part of a group of experts to analyse and report on the problems associated with meat consumption and how it relates to deforestation. Once you have completed the sections below, you will need to create a presentation to summarize your findings. You will create either a short (2-minute max.) video presentation or a presentation using Prezi/Powerpoint that includes answers to the questions below, as well as summarizing the pros and cons of deforestation. Be sure to include at least two arguments each both for and against the results of Brazilian rainforest deforestation. Use a variety of sources, reference them in your report and indicate how you know the sources are reputable.

Part 1 – Worldwide meat consumption

There has been increasing pressure on farmers to produce meat (beef, pork, lamb, chicken and other types of meat that humans eat), which has undesirable effects on the environment.

According to the Food and Agriculture Organization of the United Nations, the annual world meat consumption in 1965 was 24.2 kg per capita (per person). This had increased to 41.3 kg per capita in 2015.

a Assuming a linear model, determine the equation for worldwide meat consumption per capita since 1965.

b Use your model to predict meat consumption per capita this year.

c Predict the meat consumption per capita in 2030.

d Discuss whether you think this is a realistic model.

e Do you think this increase in meat consumption is sustainable? Explain.

To put this in perspective, take a look at the following table, which compares meat consumption in developed and developing countries.

	1997 meat consumption (million tonnes)	Average annual increase in meat consumption since 1997 (million tonnes)
Developed countries	98	0.8
Developing countries	111	4.6

f Graph each set of data on the same axes. Assume a linear model for each.

g Determine the equation of each line.

h What does the slope of each line represent? Which slope is more concerning? Explain.

i Will there ever be a time when the meat consumption of developing countries is equal to that of developed countries? Explain. If it is possible, use your equations to try to find out approximately when this will happen.

You may want to think about the populations and their growth in developed and developing countries.

Part 2 – The Brazilian rainforest

Approximately 70% of clearcutting (cutting down all or most trees in an area) in the Brazilian rainforest is to provide land for cattle ranches and farming. This accounts for almost 15% of the world's total annual deforestation, making it the largest cause of deforestation worldwide.

The graph below represents the forest cover in the Brazilian Rainforest since 1970.

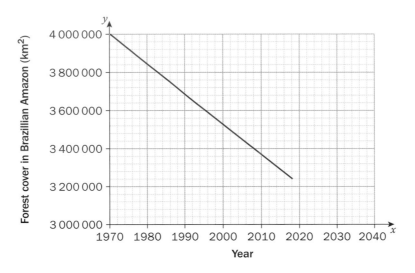

a Determine a linear equation that represents the relationship between the amount of forest cover in the Brazilian Amazon and the year. What does the y-intercept represent? What does the gradient represent?

b Assuming that the deforestation rate is constant, calculate the approximate area of forest that is cleared annually.

c To put this area in perspective, research the area of the city or town you live in in square kilometres. How does it compare with the area of the rainforest that is cleared each year?

d Given the calculated rate of area clearcut each year, how long does it take to clearcut an area of the rainforest that is the size of your town or city?

e Some scientists warn that the rainforest cannot deplete in size to less than 75% of its 1970 size without catastrophic irreversible consequences to the rainforest ecosystem. If clearcutting continues at the constant rate represented in the graph, in what year will that occur?

f Use a variety of sources and research three major reasons why we need the Brazilian rainforest. Reference your sources in your report and indicate how you know they are reputable.

Action plan

a Write an action plan containing at least five things that could be done to address the challenges of feeding a growing planet.

b Write down 3 personal decisions that you can make that will make a difference and describe their effects on the environment.

c Summarize your action plan and decisions in an interactive presentation using a program such as Prezi.

4 3D shapes

In this unit, you will see how 3D shapes have helped us to solve some complex problems with a range of truly fascinating products. But these versatile shapes can be instrumental in other contexts as well, from personal applications to ancient history.

Personal and cultural expression

Expressing beliefs and values

You are constantly surrounded by 3D shapes. Architecture, be it ancient or modern, has used the shapes studied in this unit to create some of the most inspiring and memorable structures on the planet. Many structures also have a strong cultural and personal significance.

The pyramids of Egypt are one of the Seven Wonders of the Ancient World. They were built to house the tombs of dead pharaohs, and it was believed that the shape that would allow the pharaoh to climb skyward and live forever.

Pisa was once one of four Italian republics that ruled over the Mediterranean. It was an important port and city, and it deserved a symbol of its importance and prosperity. An intricate bell tower was constructed to house the bells of the nearby cathedral. However, today it is famous for its imperfection – it was built on soft ground, and began to lean early on in its construction. Today the tower leans at about 3.99° – this could be of interest in your study of triangles too!

Civilizations and interaction

Three dimensional shapes have been an integral part of how some of the greatest peoples on Earth celebrate and interact with one another. Modern visitors to Chichen Itza are sometimes surprised to see stone goalposts surrounding a huge ballcourt. These goals were used in an ancient game played with a spherical rubber ball, not dissimilar from a soccer or basketball.

The same people who played that game also constructed the famous El Castillo pyramid at the center of the Chichen Itza site. Every spring and autumnal equinox, the Mayan people of Chichen Itza would come together for a dazzling light show as sunlight and shadows seemed to bring the serpent on the side of the temple to life.

4 3D shapes
Products, processes and solutions

KEY CONCEPT: RELATIONSHIPS

Related concepts: Generalization, Measurement

Global context:

While exploring the global context of **scientific and technical innovation** in this unit, you will examine some ingenious solutions to problems, as well as products designed to improve the quality of human life. You will take a look at both natural and man-made processes, which have produced some surprising outcomes of their own. Your study of mathematical shapes and the measurements that define them will take you on a tour of some of the more interesting results of human and natural forces.

Statement of Inquiry
Generalizing relationships between measurements can help analyze and generate products, processes and solutions.

Objectives

- Calculating the surface area and volume of 3-dimensional shapes involving cylinders, cones, pyramids and spheres

- Applying mathematical strategies to solve problems involving 3D shapes

Inquiry questions

F How do we measure space?
What is a generalization?

C How do measurements influence decisions?
How do we generalize relationships between measurements?

D What makes for an ingenious solution?
How can a product solve a problem?

ATL1 **Thinking: Creative-thinking skills**

Make guesses, ask "what if" questions and generate testable hypotheses

ATL2 **Thinking: Transfer skills**

Combine knowledge, understanding and skills to create products or solutions

You should already know how to:

1 Find the area of a circle

Find the area of the circle with the following dimensions.

a radius = 12 cm
b diameter = 10 cm
c circumference = 25 cm

2 Simplify algebraic expressions

Simplify the following expressions.

a $8r^3 + 9r^2 - 2r^2 - 11r^3$
b $3x^2(x^3)$
c $\frac{1}{2}m^4(8m^2)$

3 Solve equations

Solve the following equations. Round your answers to the nearest tenth where necessary.

a $x^2 = 14$
b $3x^2 = 25$
c $\frac{2}{3}x^3 = 10$

4 Find the volume of prisms

Find the volume of the following shapes.

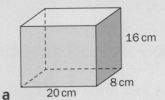

a

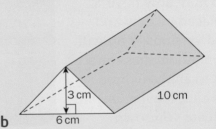

b

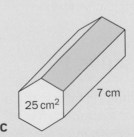

c

Introducing 3D shapes

The products of human creativity and ingenuity are all around you. Some are designed to improve our health, some to solve problems while others simply make life a little bit easier. Tools are also created to study natural phenomena so that we can understand our planet better. Some of the most amazing structures on Earth are the result of natural forces, and distinguishing between the products of human effort and natural phenomena is not always as straightforward as you might think.

For example, is the cylindrical formation in the middle of this photo natural or man-made? How might it have been formed?

In this unit, you will delve into products, processes and solutions, some of which you see every day and some of which may now have been replaced by new and improved versions. You will see how mathematics can help you to analyze these products and, maybe, inspire you to create your own.

Reflect and discuss 1

- Do you think the white cylinder in the photo on this page is the product of natural or man-made forces? Describe what you think it is and explain your reasoning.

- What product(s) do you use most often each day? What would your life be like without them?

- Name a product that you use that solves a particular problem and describe the problem that it solves.

Cylinders

Cylinders form the basis of many interesting products and processes. Knowing how to calculate the surface area and volume of a cylinder will allow you to analyze these in greater depth.

Surface area of a cylinder

Tree trunks are some of the largest naturally occuring "cylinders" on the planet. The giant sequoia in California, USA, can grow as tall as 84 m and its trunk has an average diameter of 8 m. It is from trees like this that humans have produced some incredible products such as the totem poles created by many native North American groups.

Did you know…?

The Haida Gwaii people of British Columbia solved the problem of leaky seams in canoes by carving an entire canoe out of a single Western red cedar tree. Like many First Nations people, they carved their canoes in the forest to solve the problem of having to carry a heavy tree to the water. Not all trees, however, were well suited to the task. Some of the Haida Gwaii's abandoned attempts can be found in forests to this day.

Other incredible products from trees include the totem poles created by many native North American groups.

Investigation 1 – Surface area of a cylinder

1 Using three pieces of paper and some tape, create a model of a cylinder. Use one piece of paper as the "body" or "tube" of the cylinder. Cut the other piece of paper to form the base shapes so that they fit perfectly on the top and bottom of the tube.

2 Draw the net of the cylinder, indicating important measurements on your diagram.

3 Calculate the surface area of the cylinder. Refer to your model if you are unsure about how to calculate some of the areas.

4 Draw a cylinder and label its height and radius with measurements of your choice.

5 Find the total surface area of this cylinder.

6 Generalize your work to determine a formula for finding the surface area of any cylinder if you know the measurement of its height h and and the radius r of the base.

Reflect and discuss 2

- Explain how the circumference of a circle is related to the surface area of a cylinder.

- How is finding the surface area of a cylinder different than finding the surface area of a prism? Which is easier? Explain.

Example 1

(Q) The Kake totem pole in Alaska is one of the tallest in the world. It was carved from a single tree in honor of a chief who had died. While the rings of a tree tell us about the tree's natural history, the totem poles that are carved from trees relate a human story.

Find the surface area (in m²) of the Kake totem pole that was available for carving. Assume that it has a height of 40.2 m and a diameter of 2.6 m. Round your answer to the nearest hundredth. (Assume that no carving was to be done on the bottom surface of the pole.)

▶ Continued on next page

A The available surface area of the totem pole is given by

$$SA = \pi r^2 + 2\pi rh$$

$r = 2.6 \div 2$
$r = 1.3\,\text{m}$

With $r = 1.3\,\text{m}$ and $h = 40.2\,\text{m}$, the surface area is

$$SA = \pi(1.3)^2 + 2\pi(1.3)(40.2)$$

$$SA = 333.67\,\text{m}^2$$

The formula for the surface area of a cylinder is
$SA = 2\pi r^2 + 2\pi rh$

However, the totem pole will not be carved on the bottom surface, so remove the area of one circle.

The radius of the totem pole is half of its diameter.

Substitute the measurements into the formula.

Calculate the answer and round to the nearest hundredth.

Practice 1

1 Find the surface area of the following cylinders. Round your answers to the nearest tenth.

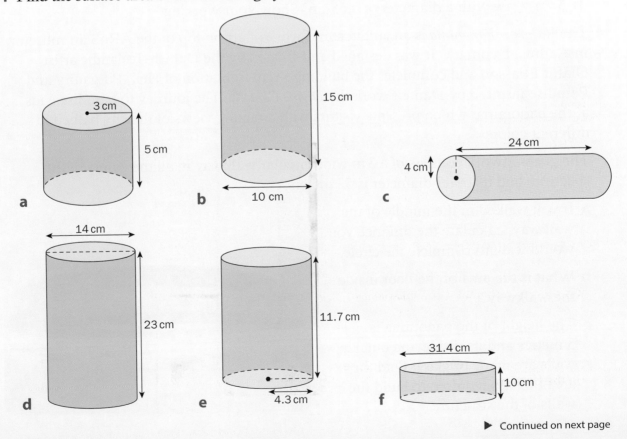

a — 3 cm, 5 cm

b — 15 cm, 10 cm

c — 24 cm, 4 cm

d — 14 cm, 23 cm

e — 11.7 cm, 4.3 cm

f — 31.4 cm, 10 cm

▶ Continued on next page

2 Copy and complete the following table for the surface area of a range of cylinders. Round answers to the nearest tenth.

Radius (cm)	Height (cm)	Surface area (cm²)
12	18	
10	11	
7		611.4
8		985
3		100

3 In the 1880s, Marvin Stone drank his mint julep using what was then the standard tool, a piece of rye grass. However, he did not like the residue caused by the grass disintegrating in the liquid. His solution was to wrap strips of paper around a pencil and glue them together, creating the first drinking straw. Most straws nowadays are made of plastic and modifications include a flexible part or even a spoon on one end. However, for environmental reasons, there is a move away from plastic and back to paper straws.

Which takes more material to make, a 20 cm straw with a radius of 0.65 cm or a 21.5 cm straw with a diameter of 0.65 cm? Show your working.

4 The *Rainbow Panorama* is an architectural work of art on top of the ARoS art museum in Aarhus, Denmark. It was designed and created by the Danish–Icelandic artist Olafur Eliasson and completes the building's representation of Hell, Purgatory and Paradise inspired by Dante's work the *Divine Comedy*. The journey to paradise ends at the panorama, which presents visitors with amazing views of the city bathed in different colors.

The glass artwork consists of a 3 m wide circular walkway in all the colors of the spectrum and its outer diameter is 52 meters.

a If you walked in the middle of the walkway, calculate the distance you would walk to complete the circle.

b What is the area of the floor inside the walkway?

c The height of the panorama is 3 meters and its inside and outside walls are made of glass. Find the area of glass required to build the walls of the structure.

5 One of the tallest trees in the world is a giant sequoia tree named "General Sherman" in California, USA. With a height of 84 m and an average diameter of 8 m, the tree can be modeled by a cylinder.

a Find the surface area of the tree that could be used to carve a totem pole. (Assume that the bottom surface will not be carved.)

b The next largest giant sequoia ("General Grant") has a height of 81.7 m and an average diameter of 7.6 m. Find the difference in surface area between General Grant and General Sherman.

c Find the height of a giant sequoia with the same surface area as General Sherman, but an average radius of 9 m.

GENERAL SHERMAN

Volume of a cylinder

As with all 3D objects, a cylinder takes up space. Its volume can be calculated in a similar way to the volume of a prism.

$V = A_{base} \times h$

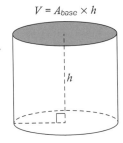

h

Activity 1 – Volume of a cylinder

1 Draw a diagram of a rectangular prism, a triangular prism and a hexagonal prism.

2 Write down the formula for the volume of each of the prisms.

3 Write down the general formula for the volume of **any** prism.

4 Draw a diagram of a cylinder. Using your result from step 3, write down the formula for the volume of a cylinder.

Reflect and discuss 3

- Explain how the volume of a cylinder is related to the area of a circle.

- If the cylinder has no top, does that affect its volume? Explain.

Example 2

(Q) When faced with the problem of trying to breathe underwater, ocean explorer Jacques Cousteau designed an underwater breathing apparatus called the Aqualung. It was the precursor of today's scuba gear and allowed Cousteau to explore places that had never been seen before. In 1972, he explored the Great Blue Hole, a cylindrical formation in Belize which began forming over 70 million years ago.

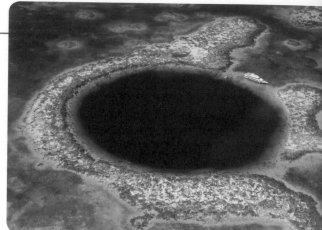

The Great Blue Hole has a depth of roughly 125 m. If there is 9 132 700 m³ of water in the Great Blue Hole, find its radius. Round your answer to the nearest tenth.

(A) $$V = \pi r^2 h$$

The formula for the volume of a cylinder is $V = \pi r^2 h$. In this case, the height of the cylinder is the depth of the Great Blue Hole.

$$9\,132\,700 = \pi r^2 (125)$$

Substitute the information that you know into the formula.

$$9\,132\,700 = (392.70)\, r^2$$

$$\frac{9\,132\,700}{392.70} = r^2$$

$$23\,256.18 = r^2$$

Solve the equation for r. Round intermediate calculations to the nearest hundredth.

$$\sqrt{23\,256.18} = \sqrt{r^2}$$

To isolate r, you need to take the square root of each side.

$$152.5 = r$$

Find the answer and round to the nearest tenth.

The radius of the Great Blue Hole is approximately 152.5 m.

Practice 2

1 Find the volume and surface area of each shape. Show all of your working and round your answers to the nearest tenth.

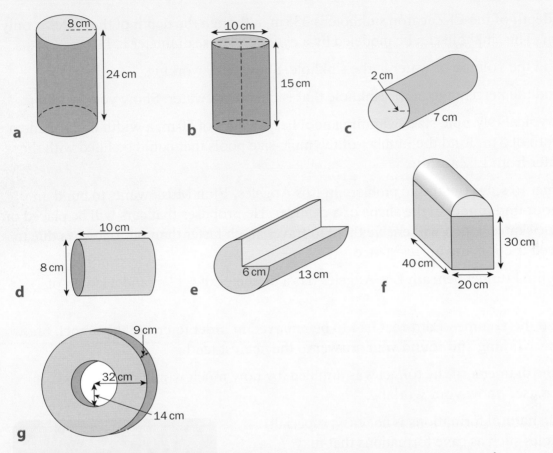

2 Find the missing dimension of each cylinder. Round your answers to the nearest hundredth where necessary.

Radius of base	Height of cylinder	Volume
4.5 cm		19.4 cm³
	15 cm	550 cm³

3 Designing tools to explore planets and moons that are very different from our own is one of the problems that NASA continuously has to solve. Often, NASA will propose a solution, build a prototype product and then test it here on Earth first. One such product is the Deep Phreatic Thermal Explorer (DEPTHX), an underwater robot that makes its own decisions on where to go, which samples to collect and even how to get back home.

▶ Continued on next page

Created to explore Europa, the smallest of Jupiter's moons, the robot was tested in the El Zacatón sinkhole in Mexico, the deepest water-filled sinkhole in the world. Sinkholes are the result of natural processes involving acidic rainwater and limestone and they can be found all over our planet.

The depth of the El Zacatón sinkhole is 335 m, although the depth of the water is only 319 m. The sinkhole can be modeled by a cylinder whose diameter is 110 m.

a Find the volume of water in the sinkhole. Show your working.

b Find the percentage of the sinkhole that is filled with water. Show your working.

c A typical Olympic-size swimming pool has a length of 50 m, a width of 25 m and a depth of 3 m. Find the number of Olympic-size pools that could be filled with the water from El Zacatón.

4 In order to solve the traffic problem in Los Angeles, Elon Musk wants to build an underground tunnel in the shape of a cylinder. He proposes that cars will be placed on electric skates which will enable them to travel much faster than on roadways due to reduced friction and air resistance.

The initial tunnel beneath Los Angeles has a diameter of 4.11 m and a length of 2.6 km.

a Find the volume of dirt that had to be removed in order to create the tunnel. Show your working and round your answer to the nearest tenth.

b If the diameter of the tunnel was doubled, by how much would the volume increase? Show your working.

5 Dating natural formations is not easy, especially examples such as cave formations that may have taken millions of years to create. However, new technologies and processes have been developed to accomplish just that. One of these processes involves measuring the amount of alunite, which is a chalky mineral that forms on cave walls as limestone is eroded from below by sulphuric acid. Another promising method uses radiocarbon dating to find the age of organic material inside the cave. The ability to accurately date caves and the formations in them could be a new tool for scientists hoping to solve the riddle of climate change over the history of our planet.

▶ Continued on next page

The Watchtower is a column in the Natural Bridge Cavern near San Antonio, Texas. The Watchtower is 15.25 m tall and has a circumference of 6.71 m.

a Find the approximate amount of material in the Watchtower, assuming there are no holes in it.

b The process of forming a column is very slow, with 16 cubic centimeters of a column being added every 100 years. (That's about the size of an ice cube every 100 years!) Find how long it took for the Watchtower to form. Show all of your working and round your answer to the nearest year.

c The surface of a cave formation is very sensitive. The oils on a person's hand can damage it forever. The owners of the Natural Bridge Cavern are thinking of covering the portion of the Watchtower that is most at risk of being touched by visitors. If they decide to use fabric up to a height of 3 meters, find the amount of fabric (in square meters) that they will they need so that it goes all the way around the Watchtower. Show all of your working.

Cones

A cone is a three-dimensional object that has a flat, circular base and a single *vertex*. When the vertex (or *apex*) is directly above the center of the base, it is called a *right cone*. In this unit, you will only study right cones, which will be referred to simply as *cones*.

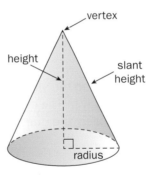

Reflect and discuss 4

- How is a cone different than a prism or cylinder? Explain.

- Why does a cone have two "heights"? Explain.

- Write down an equation relating the cone's radius, height and slant height.

Volume of a cone

Because a cone does not have a base shape that runs through the whole figure, finding its volume cannot follow the same procedure as for a prism or cylinder. However, there is a relationship between the volume of a cone and that of a cylinder, as you will see in the next investigation.

Investigation 2 – Volume of a cone

criterion B

Your teacher may give you manipulatives and sand in order to find the relationship between the volume of a cone and that of a cylinder.

If you don't have access to the materials needed, you can watch the following video demonstration: **https://www.youtube.com/watch?v=9fUmkZuaRQM** or search for "volume of a cone water" on YouTube. It uses water instead of sand, but the results are the same.

1 Look at the dimensions of the cone and the cylinder. What do you notice? Discuss your observations with a peer.

ATL1 Reflect and discuss 5

Before doing the activity or watching the video, guess how many cones full of sand it will take to completely fill the cylinder. Compare your guess with a peer. Your teacher may collect the guesses of the entire class.

2 Use the sand to determine how many cones it takes to fill the cylinder.

3 Write down the relationship between the volume of a cone and the volume of a cylinder with the same height and radius.

4 Use the formula for the volume of a cylinder and your result from step 3 to generalize a formula for the relationship between the volume of a cone and the measurement of its height h and its base radius r.

5 If you have other manipulatives, verify your relationship for another cone and cylinder with the same dimensions.

Activity 2 – The frustum

The Orion Spacecraft is currently being built with the goal of traveling deeper into our solar system than we have ever gone before. It is designed to carry up to four crew members.

The height of the crew section of the spacecraft is 3.3 m, with a large base diameter of 5 m and a smaller top diameter of 2 m. The shape of the crew module is called a frustum, which is a truncated (chopped off) cone.

1 Using the photograph and diagram as a guide, draw a diagram of the spacecraft and label its dimensions.

2 Use similar triangles to find the height of the smaller cone that has been chopped off. Hence find the total height of the cone.

3 Calculate the volume of the frustum-shaped spacecraft.

4 Only 27% of the spacecraft's volume is habitable volume that can be accessed by the crew. Calculate the habitable volume.

Reflect and discuss 6

- Estimate the dimensions of your classroom and calculate the volume. How does it compare with the habitable space available to the astronauts in the spacecraft?

- What would it be like being in that size space for an extended period of time?

- Research what NASA hopes to gain by these explorations. Would living on the spaceship in such a confined space be worth it?

Practice 3

1 Find the volume of the following shapes. Round your answers to the nearest tenth.

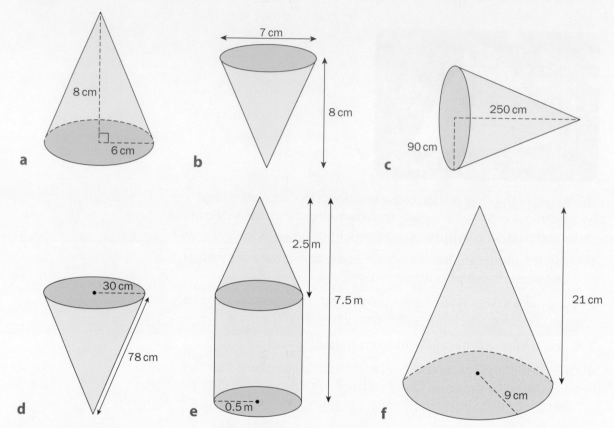

2 Italo Marchiony was an ice cream maker who sold his lemon ices on Wall Street in New York City in the early 1900s. From his pushcart, he served his product in small glass dishes but they often broke or were taken by customers. To solve this problem, he designed an edible cone-shaped dish made from a folded waffle. This was the first ice cream cone, which Marchiony went on to patent.

a If Marchiony's first cone had a base diameter of 50 mm and a height of 120 mm, what volume of lemon ice was needed to fill the cone?

b If Marchiony wanted to double the volume of the cone, but keep the same height, what base radius would he have to use? Show your working.

3 A piping bag is a product used to decorate cakes efficiently and precisely. It is cone-shaped with a small hole at the apex. Different tips are placed over the hole depending on the style of the icing decoration required. Piping bags can be made of canvas with a plastic interior or even food-grade silicone, which makes them easier to clean.

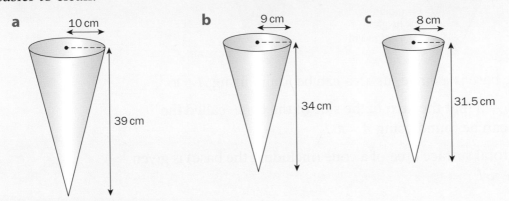

a 10 cm / 39 cm **b** 9 cm / 34 cm **c** 8 cm / 31.5 cm

Given the dimensions of each piping bag, find the volume of icing that it can contain. Round your answers to the nearest tenth.

4 Popcorn is typically served in bags or boxes. However, some companies are using "popcorn cones" to serve popcorn, especially at events where popcorn is given away for free.

a Each cone has a base diameter of 11 cm and a slant height of 20 cm. Find the volume of popcorn that will fit in the cone.

b Why would a company use a cone to serve free popcorn? What problem might that solve?

5 It has been estimated that over 1.4 billion cups of coffee are drunk worldwide every day, with American coffee drinkers drinking over three cups a day on average! Scandinavians drink, on average, about 10 or 11 kilograms of coffee per person every year! Making all of that coffee often requires filters, which are usually made out of paper. In an attempt to reduce the need for paper products, some companies have introduced reusable coffee filters.

The diagram shows a filter with a radius of 12.8 cm which can hold a total volume of 200 cm^3 of coffee.

a Find the height of the filter.

b Find the slant height of the filter.

Surface area of a cone

Finding the surface area of a cone involves opening it out into its net.

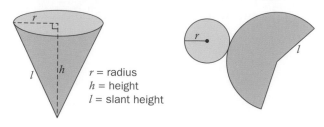

r = radius
h = height
l = slant height

Because the base is a circle, its area can be found using $A = \pi r^2$.

It can be shown that the area of the side of the cone, called the *lateral area*, can be found using $A = \pi r l$.

Hence, the total surface area of a cone (including the base) is given by $A = \pi r^2 + \pi r l$.

Example 3

Q The original ice cream cones were sold in person by street vendors, but there was a demand for the frozen treat to be available even when vendors were not in the area. Companies began producing and packaging ice creams for sale in stores.

If the diameter of each cone is 5 cm and the height is 14 cm, find the minimum amount of packaging required. Round your answer to the nearest hundredth.

A

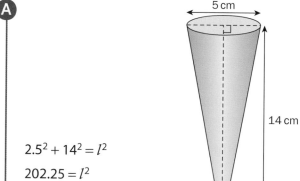

5 cm

14 cm

The formula for the surface area of a cone is $SA = \pi r^2 + \pi r l$, where l is the slant height of the cone.

$2.5^2 + 14^2 = l^2$

$202.25 = l^2$

$14.22 = l$

Find the slant height of the cone using Pythagoras' theorem.

$SA = \pi(2.5)^2 + \pi(2.5)(14.22)$

$\quad = 131.32$

Substitute the measurements into the surface area formula.

The minimum amount of packaging is 131.32 cm².

Find the answer and round to the nearest hundredth.

Practice 4

1 Find the surface area and volume of each figure. Show all of your working and round your answers to the nearest tenth.

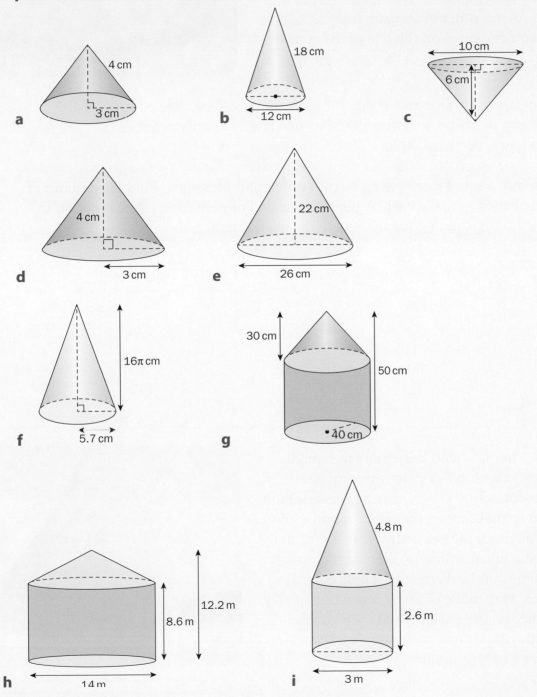

▶ Continued on next page

2 The Watercone® is a product that was invented
to help solve the problem of poor-quality
water. The circular tray at the bottom is filled
with salty water and the Watercone® is left in
the sun. As the water evaporates it rises. Salt-
free water condenses on the side of the cone
and trickles into the rim of the cone, where it is
collected.

Find the curved surface area inside the cone that could be covered in
drops at any one time, knowing that the cone has a base diameter
of 80 cm and a height of 30 cm.

3 You can buy a set of three piping bags of different dimensions. Find the amount of
material needed to make each bag (before the end of each cone shape is cut off to
make the hole). Round your answers to the nearest tenth.

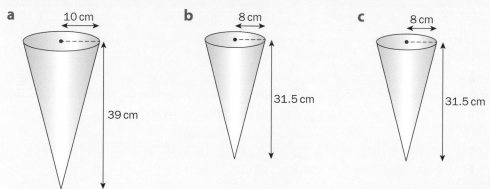

a 10 cm   39 cm

b 8 cm   31.5 cm

c 8 cm   31.5 cm

4 How do you make and keep ice in the middle
of the desert with no electricity or refrigerator?
In approximately 400 BC, ancient Persians built
yakhchāls to make ice in the colder months and
keep throughout the year. Heat rose out of the
top of the conical structure, keeping the inside
and underground cavern cold. These conical
structures were made of *sarooj*, a mixture of clay,
sand, lime, ash, egg whites and even goat hair!
The inside of the cone is empty, with the water
and ice in a cavern underneath it.

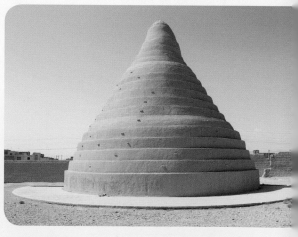

Find the approximate surface area of a *yakhchāl* whose height is
18 m and base area is 710 m².

5 The Elizabethan collar, sometimes referred to as the "cone of shame", is one solution to the problem of pets trying to lick injured areas after a medical procedure. The shape is called a *truncated cone* because it is a cone whose apex has been truncated (removed). A typical Elizabethan collar is made of plastic and is attached to the pet's collar.

If the dimensions of an Elizabethan collar are as shown in the diagram, find the area of plastic required to make it.

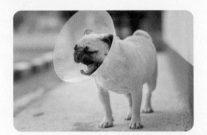

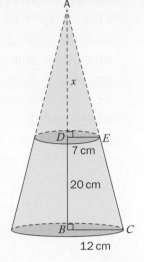

Formative assessment – Solar energy 1

criterion D

Solar energy is becoming increasingly important as one potential solution for the planet's energy needs. However, transforming the energy of the sun into electricity requires the use of photovoltaic cells, which must be mounted in areas that receive a lot of sunlight. Mounting these cells has become a problem on its own, and a wide range of solutions have been proposed. In this task, you will analyze products that have been created to try to improve the effectiveness of solar power. Please show your working throughout the task.

Cylindrical solar cells

The traditional solar panel is a rectangle. However, manufacturers of the cylindrical solar panel claim that it is a much more efficient way of collecting solar energy, especially on larger buildings. The cylinder shape allows absorption of the Sun's rays from just about every angle as it passes overhead. (A traditional panel is most efficient when the Sun is directly overhead.) When mounted on a roof that has been painted white (such as many businesses have today), even the underside of the panel will absorb sunlight that is reflected off the roof. The shape also reduces wind resistance, which makes these solar cells easier and less expensive to install.

The height of each cylinder that is covered by photovoltaic cells is 100 cm. The diameter of each cylinder is 22 mm.

a Find the total surface area that is exposed to the Sun if 200 of these solar cells are installed on the roof of a building.

b Find the amount of space (volume) that the 200 cells take up.

Spin cell cones

Spin cell cones use some of the solar energy that they generate to rotate, but their manufacturer explains that the rotation allows them to collect the Sun's rays from every angle without overheating. Most of the cone is covered in photovoltaic cells, with the exception of its base.

c If the radius of one cone is 0.55 m and its height is 0.82 m, find the area of the cone covered in solar cells.

d Find the volume occupied by an array of 50 of these spin cell cones.

ATL2 **Making choices**

You have a section of a flat white roof with dimensions of 10 m by 6 m.

e A typical rectangular solar panel measures 1.65 m by 1 m. Sketch the configuration of these panels that will cover the maximum possible area of the roof. Calculate the area that they will cover.

f Cylindrical solar cells are installed side by side and need at least 20 mm between them. Using the dimensions given above, sketch the configuration that will cover the maximum possible area of the roof. Calculate the total surface area of these cylindrical panels that will be exposed to the Sun.

g Spin cell cones can be installed so that they touch each other. Sketch a configuration of spin cell cones that will cover the maximum possible area of the roof. How many cones will you be able to place on the roof? Calculate the total surface area of the cones that will be exposed to the Sun.

h What factor(s) would influence your choice of solar panels to install? Explain.

> The area that is covered in photovoltaic cells does not include the top and base of the cylinder, so do not include these in your calculation.

Pyramids

A pyramid is another shape that meets at an apex. However, unlike a cone with its circular base, a pyramid has a base that is a polygon. All of the sides of a pyramid are triangular and each type of pyramid is named after the shape of its base. There are square pyramids, triangular pyramids, hexagonal pyramids, etc. Relationships between the various measurements (height, length, apothem) can help you to determine the surface area and volume of any pyramid.

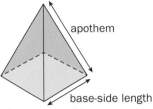

apothem

base-side length

Square pyramid

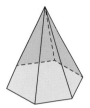

Triangular pyramid

Hexagonal pyramid

Did you know...?

A triangular pyramid is often referred to as a *tetrahedron*. Some molecules in chemistry, like the methane molecule shown here, are tetrahedrons. The central atom is located at the center of the pyramid (carbon, in this case), while the four surrounding atoms (hydrogen) occupy the vertices.

Volume of a pyramid

In the same way as you did for the cone, you can find out the volume of a pyramid by carrying out an investigation into its relationship to the volume of a prism with similar dimensions.

Investigation 3 – Volume of a pyramid

Your teacher may give you manipulatives and sand in order to find the relationship between the volume of a rectangular pyramid and that of a prism with the same height and base area.

criterion
B

If you don't have access to the materials needed, you can watch the following video demonstration: **https://www.youtube.com/watch?v=O2wenAIf0H8** or search for "volume of a pyramid water" on YouTube.

▶ Continued on next page

1 Look at the dimensions of the prism and the pyramid with the same base. What do you notice? Discuss your findings with a peer.

Reflect and discuss 7

- Before doing the activity or watching the video, guess how many pyramids full of sand it will take to completely fill the prism. Compare your guess with a peer. Your teacher may collect the guesses of the entire class.

- Will your answer depend on the type of prism and pyramid? Will a triangular pyramid fill a triangular prism in the same way as a rectangular pyramid and rectangular prism? Explain.

2 Use the sand to determine how many square pyramids it takes to fill the square prism.

3 Repeat step 2 with other pyramids and their corresponding prisms if those shapes are available.

4 Write down the relationship between the volume of a pyramid and the volume of a prism with the same base and height. Write this relationship as a formula.

5 Use the formula for the volume of a prism and your result from step 4 to write down the formula for the volume of a square pyramid with height h and base with measurements l and w.

6 If possible, verify your relationship for another pyramid and prism with the same dimensions.

Reflect and discuss 8

- How does your result in this investigation relate to your result in Investigation 2 (the investigation into the relationship between the volumes of cones and cylinders)? Explain.

- Generalize the formula for the volume of any pyramid, regardless of its base shape.

Practice 5

1 Find the volume of the following figures. Round answers to 3 s.f.

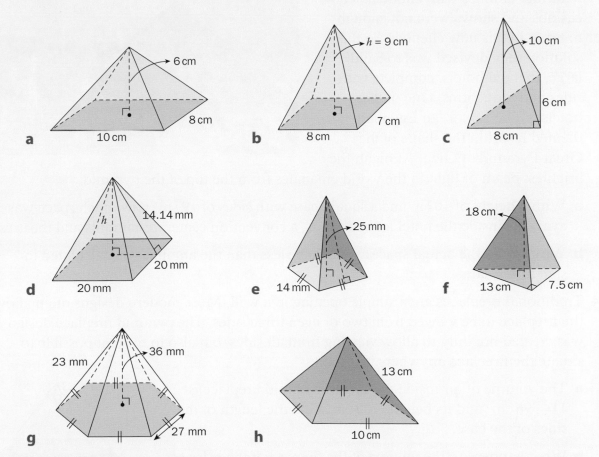

2 Find the missing measurements in each of the following square pyramids.

Height (cm)	Base side length (cm)	Apothem (cm)	Volume (cm³)
4	6		
	16	10	
12		13	
	20		1000
8			314
3.25			702.15

> If you are finding calculating this difficult, try adding an extra column for the base area.

▶ Continued on next page

3 In the 1990s, hotels on the Las Vegas Strip began a ten-year project to try to attract families with children. The casinos and shows were not enough to bring in this new clientele, so the solution they devised was to build large, themed hotels, complete with rides and attractions. One of these hotels is The Luxor, an Egyptian-themed resort in the shape of the Great Pyramid of Giza. At night, the brightest beam of light in the world emanates from the top of the pyramid.

a With a height of 107 m and a square base with sides of 197 m, how much space was available inside the hotel for bedrooms, a convention center, restaurants and theatres?

b Why would the actual space available be less than the amount you calculated in step **a**? Explain.

4 Traditional fireplaces are a simple opening in a wall. More modern designs often allow the fireplace to be viewed from two or even three sides. The pyramid fireplace design was created not only to allow viewing from all sides, but also to make it possible to situate the fireplace anywhere in a room.

a The volume of air inside the square pyramid fireplace is 114 996 cm^3 and its height is 63 cm. Find the length of the sides of the base.

b What happens to the volume of the fireplace if the sides of the base are doubled in length? Show your working.

Surface area of a pyramid

The entrance to the Louvre museum in Paris is a giant pyramid made of glass that is a scale model of the Great Pyramid of Giza in Cairo, Egypt. Completed in 1989, the structure was criticized by many, although it is now much more accepted and even admired. The pyramid was designed to help solve the problem of accessing the Louvre's collections since the original entrance could not handle the large volume of visitors to the museum. Using a glass pyramid also meant that the underground lobby would be well lit while preserving the view across the courtyard

from any point outside the pyramid. It also paid homage to the extensive Egyptian collection within the museum.

How much glass do you think was needed to make the Louvre Pyramid? To answer this, you need to find out how to calculate the surface area of a pyramid.

Activity 3 – Surface area of pyramids

Creating a net is an efficient way to determine the surface area of a 3D shape, since you can flatten the shape to see all of its faces and then add the areas of the individual sections.

1 Copy and complete this table.

3D solid	Dimensions and shapes of sections needed to determine surface area (SA)	SA formula
Triangular pyramid		
Rectangular pyramid		
Square pyramid		
Hexagonal pyramid		

Pairs

2 In pairs, follow these steps to determine the formula for each pyramid listed in the table.

- Draw the net of each 3D shape in the table.
- Label the dimensions of each section.
- Find the area of each section.
- Total the areas of all of the sections.
- Use this to write the formula for the surface area of the prism.

Your teacher may provide you with manipulatives of 3D pyramids that can be broken down into their nets or, alternatively, you can look at virtual manipulatives.

Search online for 'Annenberg Interactives Pyramids' to see how the net unfolds. Use this as a visual to help you determine the formulas.

Example 4

Q The Louvre Pyramid is a square pyramid whose base has sides measuring 35.4 m and whose apex is 21.6 m above the ground. Find the amount of glass needed to cover the sides of the pyramid in square meters. Round your answer to the nearest square meter.

A

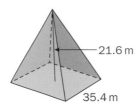

21.6 m

35.4 m

Draw the net of the 3D figure and indicate the known measurements.

35.4 m

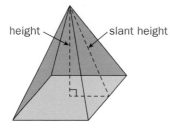

height   slant height

To find the surface area of each triangle, use Pythagoras' theorem to calculate the slant height (s).

$s^2 = h^2 + 1^2$

$s^2 = 21.6^2 + 17.7^2$

$s^2 = 779.85$

Round calculations to the nearest hundredth in the middle of a solution.

$s = 27.93$

$A = \dfrac{1}{2}(35.4)(27.93)$

Find the area of one triangular face and multiply by four.

$A = 494.36$

The area of all 4 triangles $= 4(494.36)$

$= 1977.44$

You do not need to find the area of the base since there is no glass on the base.

$SA = 1977 \text{ m}^2$

Round the answer to the nearest square meter (as specified in the question).

Practice 6

1 Find the volume and surface area of the following figures.
Round answers to the nearest tenth.

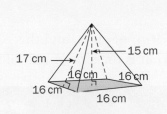

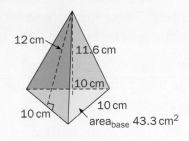

a

b

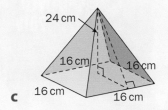

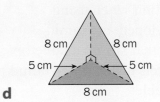

c

d

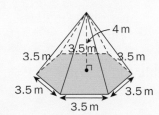

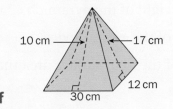

e

f

2 Dice play an important part of role-playing
games. Because not all scenarios have six
options, it was necessary to make dice with
different numbers of sides. You can now get dice
with 4, 6, 8, 12 and even 20 sides!

A four-sided die (abbreviated d4) is a triangular
pyramid, all of whose faces are equilateral
triangles.

a If the apothem of any side of the pyramid
is 18 mm and the length of the sides of the
triangles is 10.4 mm, find the surface area of the dice.

b Is a d4 fair? Explain.

▶ Continued on next page

3 Many of the collections at the Louvre can be accessed underground via hallways and corridors. The designer of the Louvre Pyramid wanted to use as much natural light as possible to illuminate these subterranean passageways. In addition to the main Louvre Pyramid, there are also three smaller ones that serve as skylights that let in light where three main hallways meet. These, too, are square pyramids and are made of glass (with the exception of the base).

If each of the three small pyramids is covered in a total of $79\,m^2$ of glass, and the apothem of each of the triangular sides measures $6.75\,m$, what is the side length of the base of each small pyramid?

4 Using technology is often seen as an individual activity. One company hopes to solve this problem by promoting group interaction when using technology. They have developed a pyramid-shaped holographic display that sits on a tabletop. Users can interact with objects inside the pyramid using their smartphone or tablet.

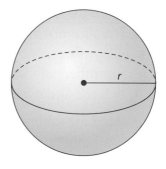

The device is a square pyramid with sides measuring $45\,cm$. The height of the pyramid is $23\,cm$.

a Find the surface area of the entire device.

b Find the percentage of the total surface area that is available for viewing.

Spheres

A sphere is a perfectly round three-dimensional object. In mathematics, it is often defined as "the set of all points in three dimensions that are the same distance from a fixed point".

Reflect and discuss 9

- Explain how a sphere represents the set of points in three dimensions that are the same distance away from a fixed point.
- Is the sphere the entire solid "ball" or just the shell? Explain.
- What is the "fixed point" called?

Surface area of a sphere

Unlike prisms, cylinders and pyramids, a sphere cannot be broken down into individual faces nor can you draw a net of a sphere accurately. How then can you determine its surface area?

Investigation 4 – Surface area of a sphere

criterion **B**

For this investigation, you will need an orange, a tangerine or a clementine. Choose one whose shape is as spherical as possible!

If you don't have an orange available, you can watch the following video demonstration: **https://www.youtube.com/watch?v=jaL8Kuv6YHo** or search for "surface area of sphere orange" on YouTube.

1 Place the orange on a piece of paper. Holding your pen or pencil upright, trace around the orange to make a shape that is roughly circular.

2 Repeat step 1 several times so that you have between 4 and 6 circles that are roughly equal in size.

3 In the middle of each circle, write the formula for the area of the circle, assuming its radius is r.

4 Peel the orange in small pieces. Try to make the pieces no larger than the fingernail of your thumb.

ATL1 ## Reflect and discuss 10

Before doing the activity or watching the video, guess how many circles you will completely fill with orange peel. Compare your guess with a peer. Your teacher may collect the guesses of the entire class.

5 Completely fill each circle before moving on to the next. Try to avoid leaving any space between pieces. You may not fill every circle.

6 How many circles did you fill? What is the total area of those circles in terms of r?

7 Collect data from several of your peers. How many circles did each person fill?

8 Based on your result, write down the formula for the surface area of a sphere.

- Why does the amount of orange peel represent the surface area of a sphere? Explain.

- What other objects could you use to perform this investigation? Explain.

- What factors may have contributed to students in your class filling more or less than the correct number of circles? Explain.

Example 5

Q Throbber heating balls are designed to heat food, and can be controlled using a smartphone app. The balls are made of aluminium and use induction to heat up any liquid.

Suppose six heating balls are placed in a pan of soup. If the balls have a diameter of 5 cm, find the total area of the heating balls that is in direct contact with the soup. Round your answer to three significant figures.

A $r = \dfrac{5}{2}$

$= 2.5 \, \text{cm}$

> Find the radius of each ball by dividing the diameter by 2.

$SA = 4\pi r^2$

> Find the surface area of one ball using the formula $SA = 4\pi r^2$.

$= 4\pi(2.5)^2$

$= 25\pi \, \text{cm}^2$

> Keep your working in terms of π at this stage, so you don't have to round your answer until the very end of the question.

Total surface area $= 6 \times 25\pi \, \text{cm}^2$

> Since there are 6 heating balls, multiply the surface area of one ball by 6.

$= 150\pi \, \text{cm}^2$

$= 471 \, \text{cm}^2$

> Calculate the value and round your answer to 3 s.f.

The total surface area of the 6 heating balls is 471 cm².

Practice 7

1 Find the surface area of the following shapes. Round answers to the nearest tenth.

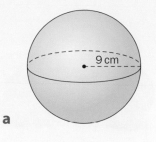

a

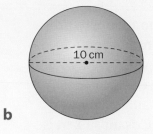

b

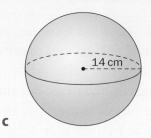

c

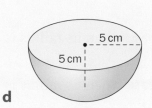

d

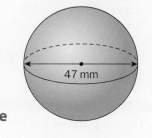

e

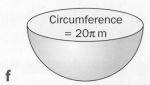

f

2 With the recent advances in portable camera technology, it was only a matter of time before someone invented a spherical camera. These ball cameras are designed to take panoramic photos when thrown into the air. With cameras mounted all over its surface, the ball camera is a new tool that amateur photographers can use to take interesting shots.

If there is 1520 cm² of surface area on the ball available for mounting small cameras, what is the diameter of a ball camera?

3 The mirrored ball was created in order to reflect light around the inside of night clubs. The sphere is covered with small mirrors and rotates while a light shines onto it. The small mirrors reflect light into different parts of the club as the ball turns. Used as early as the 1920s, the balls gained wide popularity in 1970s disco clubs and are often referred to as "disco balls".

If the typical disco ball has a radius of 42 cm, what surface area can be covered with mirrors? Round your answer to the nearest hundredth.

► Continued on next page

4 Which would be more effective at heating up a liquid, 10 Throbber heating balls, each with a radius of 6 cm or 5 heating balls each with a radius of 12 cm? Show your working and justify your answer.

5 In 1963, Aquilino Cosani, an Italian manufacturer of plastics, perfected a process to produce puncture-resistant plastic balls. At first, they were used with infants and newborns, but they have since been used for physical therapy, exercise and even as an alternative to a regular desk chair. Cosani's exercise balls come in a wide range of sizes, from 22 cm in diameter to 85 cm in diameter.

   a By what factor does the surface area of an exercise ball increase when its radius is doubled? Use mathematics to justify your answer.

   b By what factor does the surface area of an exercise ball increase when its radius is tripled? Use mathematics to justify your answer.

   c Generalize the relationship between an increase in radius and its effect on the surface area of a sphere.

Volume of a sphere

Since there is no base shape, how do you find the area of a sphere? Like the cone, the volume of a sphere is related to that of a cylinder.

Reflect and discuss 12

- How do you find the volume of a regular prism or a cylinder? What principle is it based on?

- Can this principle be used to find the volume of a sphere? Explain.

Because a sphere does not have a base shape, its volume will have to be calculated using a different method than the standard one for prisms and cylinders.

Investigation 5 – Volume of a sphere

For this investigation, your teacher may provide you with manipulatives and sand.

If manipulatives are not available, you can watch the following video demonstration: **https://www.youtube.com/watch?v=8jygxFuLoCk** or search for "volume of a cylinder and sphere water" on YouTube.

1 If you have manipulatives, fill the sphere with sand.

Reflect and discuss 13

Before doing the activity or watching the video, guess the volume of the sphere as a fraction of the volume of the cylinder. (Assume that the sphere and cylinder have equal radii and that the height of the cylinder is equal to the diameter of the sphere.) Compare your guess with a peer. Your teacher may collect the guesses of the entire class.

2 Determine what fraction of the cylinder's volume is the volume of the sphere by pouring sand from the sphere into the cylinder.

3 Write down the formula for the volume of a cylinder. Based on your work in step 2, adapt this formula to write down a formula for the volume of a sphere of equal height and radius.

4 What is relationship between the height of the cylinder and the radius of the sphere?

5 Write down the height of the cylinder in terms of the radius of the sphere. Substitute this into your formula from step 3.

6 Simplify your formula, so that you end up with the formula for the volume of a sphere.

Reflect and discuss 14

- How close was your initial guess of the relationship between the volume of the sphere and the cylinder?

- How does this result relate to the volume of other shapes that you have studied in this unit?

Example 6

Q The Super Ball was one of the many toy fads of the 1960s and 1970s. It was a bouncing ball made of a material called Zectron and it was said that if it was thrown hard enough onto the ground it could bounce higher than a three-story building! One SuperBall contains 451.76 cm³ of Zectron. Find the radius of the ball. Round your answer to the nearest hundredth.

A

$$V = \frac{4}{3}\pi r^3$$

$$451.76 = \frac{4}{3}\pi r^3$$

$$\frac{3}{4}(451.76) = \frac{3}{4}\left(\frac{4}{3}\pi r^3\right)$$

$$338.82 = \pi r^3$$

$$\frac{338.82}{\pi} = \frac{\pi r^3}{\pi}$$

$$107.849756 = r^3$$

$$\sqrt[3]{107.849756} = \sqrt[3]{r^3}$$

$$r = 4.76\,\text{cm}$$

| The formula for the volume of a sphere is $V=\frac{4}{3}\pi r^3$ |

| Substitute the given value into the formula. |

| Solve for r by first multiplying each side by the reciprocal of $\frac{4}{3}$ and simplifying. |

| Divide each side by π and simplify. |

| Take the cube root of each side. |

| Round your answer to the nearest hundredth. |

Only the final answer should be rounded to the nearest hundredth. Intermediate calculations should use more significant figures in order for the answer to be as exact as possible.

The radius of the SuperBall is 4.76 cm.

Practice 8

1 Find the volume and surface area of the following shapes. Round your answers to the nearest tenth.

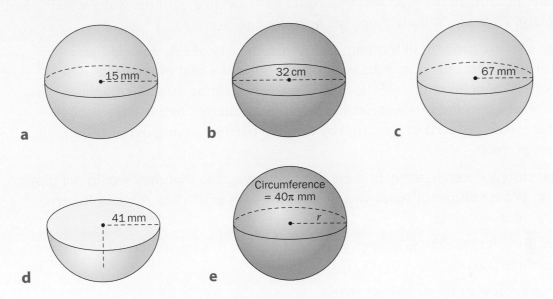

a 15 mm

b 32 cm

c 67 mm

d 41 mm

e Circumference = 40π mm, *r*

2 Copy and complete this table, assuming that each set of measurements are for a sphere. Round all values to the nearest tenth.

Radius (cm)	Diameter (cm)	Surface area (cm²)	Volume (cm³)
15			
	50		
		1000	
			1000

3 The water walking ball (or aqua zorbing ball) was invented to allow people to "walk" on water. It is made of plastic and the ball is zipped closed with the user inside.

a If the surface area of a water walking ball is $16\pi \, m^2$, find the volume of air inside the ball.

▶ Continued on next page

189

b In order for an adolescent to be able to breathe enough oxygen for 30 minutes, they need $2\,m^3$ of fresh air. Find the diameter of a water walking ball that would allow an adolescent to remain in it for 30 minutes uninterrupted.

ATL2 **c** What are some of the problems that could be associated with the water walking ball? Explain. How could these problems be overcome?

4 During a severe drought in southern California, the city of Los Angeles used small black spheres to try to reduce the amount of water that evaporated from its reservoirs. About 96 million black spheres, each measuring 10 cm in diameter, were released into the Los Angeles Reservoir in an attempt to block the sunlight. The shade balls reduced evaporation by 85 to 90 percent, saving over 1 trillion liters of water per year.

a Interestingly enough, the balls were filled with water so that they would not blow away. What volume of water was in all of the balls in the Los Angeles Reservoir? Show your working.

b Why do you think black spheres were used instead of some other shape or color? Give three reasons.

5 Find the volume of the following shapes.

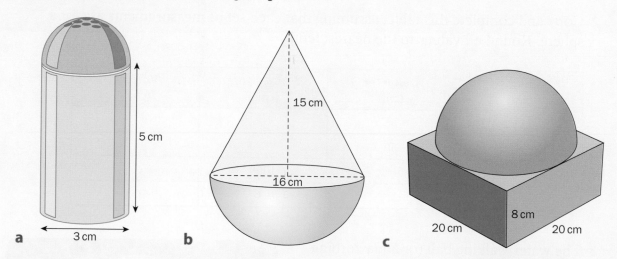

a   3 cm   b   c

5 cm, 15 cm, 16 cm, 20 cm, 8 cm, 20 cm

6 A soccer ball is packaged in a cubic box with side lengths of 22 cm. Assuming that the ball touches all faces of the box, what percentage of the volume of the box is wasted space?

7 Find the radius of a sphere that has an equal surface area and volume.

Formative assessment – Solar energy 2

You have already seen how solar cells have been made in the form of cylinders and cones, but other designs have also been attempted.

criterion **D**

Solar pyramid

The solar pyramid was designed to help solve the problem of the solar cells overheating. This device is made for much smaller applications than the cylindrical cell, such as camping or charging tools at a construction site. The square base measures 25 cm on each side and has a height of 35 cm.

a Assuming the pyramid sits on the ground, find the surface area available for solar cells. (Assume that they can be placed all the way to the top.)

b If each pyramid is filled with air, find the volume of air inside each pyramid.

Rawlemon

Inspired by his daughter's toy marbles, German architect Andre Brossel created the Rawlemon. It combines the benefits of a magnifying glass with the potential of solar energy in order to produce up to 70% more energy than a typical photovoltaic panel.

Rawlemon works with both sunlight and moonlight and is capable of following the movement of the Sun or Moon thanks to its motorized base. The large crystal ball is filled with water to help magnify the light rays on the solar cells.

The Rawlemon comes in a variety of sizes.

c Find the area of the sphere that is exposed to sunlight if it has a diameter of 1.8 m. (Assume that the whole of the sphere's surface area is exposed to sunlight.)

d Find the volume of water in a sphere with a diameter of 1.8 m.

 ATL2

Making choices

e Try out several different shapes, then choose one to design your own product on which solar cells would be placed.

f Find the surface area available for solar cells, rounded to the nearest tenth.

g Find the volume of space that your design occupies, rounded to the nearest tenth.

h Explain the usefulness of your product. What is the most appropriate use for your design (e.g on a building, for charging a small device)? Who would want to use it?

Unit summary

The volume of any 3D shape can be calculated by taking the area of the *base figure* and multiplying by the height. The base figure is the figure that has been repeated in order to produce the 3D shape.

Shape		Surface area	Volume
Cylinder		$2\pi r^2 + 2\pi rh$	$\pi r^2 h$
Cone		$\pi r^2 + \pi rs$	$\frac{1}{3}\pi r^2 h$
Square pyramid		$b^2 + 2bs$	$\frac{1}{3}b^2 h$
Rectangular pyramid		$lw + 2bs$	$\frac{1}{3}lwh$
Sphere		$4\pi r^2$	$\frac{4}{3}\pi r^3$

The surface area of any pyramid can be calculated using its net.

The volume of a pyramid is $\frac{1}{3}$ the volume of the prism with the same base shape and same height.

Unit review

criterion **A**

📄 **Launch additional digital resources for this chapter**

Key to Unit review question levels:

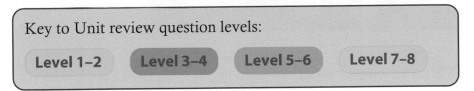

Level 1–2   **Level 3–4**   **Level 5–6**   **Level 7–8**

---

1 **Calculate** the volume and surface area of each of the following shapes.
Round answers to the nearest tenth.

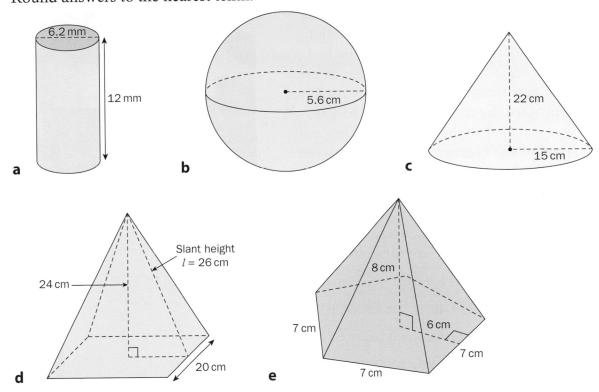

a — cylinder: 6.2 mm (diameter), 12 mm (height)

b — sphere: 5.6 cm

c — cone: 22 cm, 15 cm

d — pyramid: Slant height $l = 26$ cm, 24 cm, 20 cm

e — pyramid: 8 cm, 7 cm, 6 cm, 7 cm, 7 cm

2 In order to help reduce employees' stress levels and give them a place to get away from their office cubicle, Amazon has built three giant spheres in downtown Seattle, USA. Inside the spheres, the climate is very similar to that of Costa Rica, providing a nice break from the sometimes cold and rainy Seattle weather. The spheres also contain waterfalls, a river and tropical gardens. The largest sphere measures 40 m in diameter.

a Find the volume of space available in the largest sphere. Round your answer to the nearest tenth.

b If the spheres are covered in glass, find the area of glass needed to cover the largest sphere.

3 Airbags became widely used in cars in the 1980s and 1990s as a method of making them safer. Although seat belts had become a standard safety feature, injuries still persisted as the driver and passengers impacted the inside of the car. Airbags are installed so that, in a crash, they will deploy and create a soft barrier between the driver/passenger and the car and its windows. A car can have multiple airbags, each positioned to protect a specific passenger. The driver's airbag is located in the steering wheel of the car, and when fully inflated can be approximated by a cylinder.

If the cylinder has a height of 25 cm and a radius of 28 cm, find:

a the volume of air in the airbag when fully inflated

b the amount of material required to make the airbag.

4 In order to solve problems of manoeuvrability and stability, inventor James Dyson inserted a ball into his vacuum cleaner design. The motor could be moved to fit inside the ball, which lowered the center of gravity and increased the stability of the machine. At the same time, the ball allows the machine to be turned around corners with ease.

If the motor for a particular vacuum cleaner has a volume of 4200 cm³, find the radius of the smallest ball that could be used to hold it.

5 Three-dimensional puzzles are popular toys for adolescents, especially since the Rubik's cube. The wooden puzzle shown here has only 10 pieces but is very difficult to solve. The puzzle comes with the pieces arranged in the shape of a cube, and the aim is to rearrange them to make a square pyramid. If the puzzle contains 9 cm³ of wood, find the dimensions of the pyramid.

6 The LifeStraw personal water filter removes 99.9999% of waterborne bacteria. It can turn up to 1000 liters of contaminated water into safe drinking water. If the length of the straw is 9 inches and the diameter is 1 inch, **calculate** the volume of water that the straw can hold.

7 New Zealand, which is known for its adventurous activities, is the home of "zorbing". This involves rolling down a hill in an inflatable ball. The zorb is actually a sphere within a sphere, with a layer of air in between to help absorb shocks. Zorbs can hold up to three people and zorbers can even choose the "aqua option", where the inner sphere is partly filled with water. This makes for a refreshing ride down the hill on a hot day.

If the outer sphere has a diameter of 3 m and the inner sphere has a diameter of 2 m, find the volume of the air layer between the inner and outer spheres. (The plastic material for each sphere is approximately 0.8 mm thick, which can be ignored in these calculations.) Round your answer to 3 s.f.

8 Find the volume and surface area of each of the following 3D shapes.

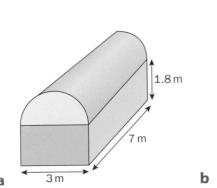

a 1.8 m 7 m 3 m

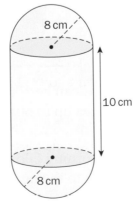

b 8 cm 10 cm 8 cm

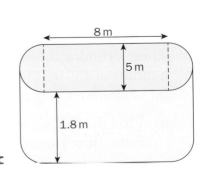

c 8 m 5 m 1.8 m

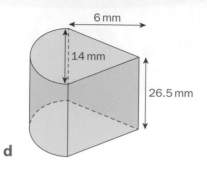

6 mm

14 mm

26.5 mm

d

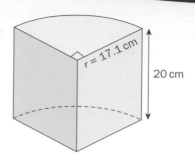

r = 17.1 cm

20 cm

e

9 Specialized paint has been designed to protect storage silos against corrosion and to prolong their life. First the silo is painted with a layer of rust-resistant primers and then with two layers of the specialized paint.

The height of the cylindrical part of one agricultural silo is 17 m and the total height of the silo is 20 m. The diameter of the silo is 15 m. Find out how much paint and primer are needed to protect the five silos shown here.

10 The Morning Glory cloud formation (shown in the photo) occurs very predictably on the north coast of Australia. Although its causes are not known precisely, it forms mostly in October over the Gulf of Carpentaria during a process when sea breezes from the west meet breezes from the east.

To study such phenomena, scientists use a tool called a radiosonde, attached to a weather balloon (which is spherical!).

These cylindrical cloud formations can be 1000 km long with a radius of 150 m. Moving at speeds up to 60 km/h, there is a lot of cloud moving overhead.

a Find the surface area of the Morning Glory cloud formation. Round your answer to the nearest hundredth.

b If the cloud formation were only half as long, by what factor would that change the surface area? **Justify** your answer with calculations.

c If the cloud formation had only half the radius, by what factor would that change the surface area? **Justify** your answer with calculations.

11 A common product found in many restaurants and kitchens is the pepper grinder. It allows users to grind their own pepper, keeping it as fresh as possible until used. Pepper grinders can come in a wide range of shapes and sizes.

One product is in the shape of a cylinder with a height of 14 cm and a radius of 5 cm. Another is in shape of a ball (sphere). Find the radius of a ball pepper grinder if it has to contain the same volume of pepper as the cylinder.

12 Access to clean water is a problem in many developing countries, especially as clean water sources are often located far away from the people who need the water. The Q Drum water transporter was developed to solve this problem. Its design allows it to transport a large volume of water over long distances simply by rolling the drum. This means that anyone can use the device, including children. The Q Drum holds 50 000 cm³ of water. Its height is 36 cm and the diameter of the base is 50 cm. Find the radius of the inner hole through the center of the drum.

13 The traditional igloo, with its approximately hemispherical dome, is traditionally built by Canadian Inuit and the people of Greenland. It is made of compacted snow, which is a poor conductor of heat, so any heat generated inside the structure will stay inside.

a If the radius inside an igloo is 1.8 m, find the volume of air contained inside.

b If the thickness of the walls of compacted snow is 0.5 m and the inner radius of the entrance is 0.8 m, find the surface area of the outside of the igloo.

> Disregard the entranceway in your answers to this question. Just find the volume and surface area of the hemisphere.

14 The two-piece hard capsule, patented in 1847, was designed both to ensure the correct dosage of a medication and to make the medicine easy to swallow, without having to taste the medicine. The pharmaceutical manufacturer fills the hollow gelatin capsule with medicines such as antibiotics.

Capsules come in many sizes, but a common one has a total length of 11 mm and a diameter of 5 mm.

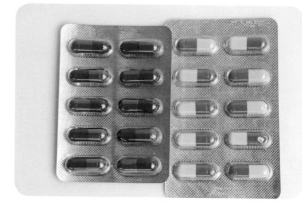

a Find the volume of medicine that a capsule of this size can hold. Include a diagram with your calculations, showing the dimensions of the capsule.

b Find the amount of gelatin needed to produce one capsule.

Summative assessment

criteria
C, D

Designing a heat bag

Heating pads are commonly used worldwide to help to relax muscles and relieve pain, such as arthritis, headaches, strains, sprains, and for long-term use to alleviate chronic back pain and joint injuries. They can also be used simply to warm you up, for example when you get into a cold bed.

Your task is to design a microwavable heat bag with a volume of $1500\,cm^3$. The bag will be completely filled with a filler such as wheat, flax or other grain that can be microwaved. The longer the bag maintains its heat the better, so you may want to experiment with several different kinds of filler, or even a combination.

The bag needs to fit inside a microwave oven, so it cannot be larger than the circular glass dish found in most microwaves which has a diameter of 30 cm. The maximum height of the bag so it will not touch the roof of the microwave is 20 cm.

Part 1 – The product: possible basic shapes

a Design five possible heat bags that have the given volume using each of the following shapes.

- Square-based rectangular prism
- Cylinder
- Cone
- Square-based pyramid
- Sphere

b Draw a diagram of each design and calculate the amount of material you would need to make each (the surface area). Show all of your working. Think about what degree of accuracy you will use (how many decimal places you will round to).

Part 2 – The product: a creative shape

a Design a heat bag that has the given volume but is a compound shape made up of two or more of the above shapes put together

b Draw a diagram of your design and calculate the amount of material you will need to make it. Show all of your working. Think about what degree of accuracy you will use.

c Show that your chosen dimensions produce a 3D shape with a volume of $1500\,cm^3$.

d Choose a specific market for your heat bag (e.g. children, senior citizens). Tailor your design for this market – you can use decorations, embellishments, scents, etc. Add these design elements to your diagram.

e A common width of a bolt of fabric is 110 cm. Construct a diagram showing how you will position the net of your heat bag on a piece of fabric 110 cm wide. You should use the minimum length of fabric possible in order to minimize waste of fabric. Make sure your net is to scale.

Part 3 – Marketing materials

You will create marketing materials for your product to show its development from beginning to end as well as how you plan to sell the product. You will create a folder (digital or hard copy) that includes the following:

- the five initial designs, complete with diagrams, calculations and specifications (dimensions, surface area)

- an analysis of each of these designs, including advantages and disadvantages of each

- your final compound design, which could be created using Tinkercad, complete with calculations and specifications (dimensions and surface area)

- a sketch of the net for the compound design, drawn to scale and showing how it will best fit onto the fabric
- a justification of the degree of accuracy used given the context of the problem
- an explanation of why you chose your finished compound design and how your suggested modifications/additions make it appealing for your chosen market.
- promotional material, which can be in the form of a print or video advertisement for the product. (Online tools like Canva or Animoto can be used for this purpose.) Be sure it is geared towards your target market.

Extension

The longer the heat bag maintains its heat the better, so your choice of filler is important, but there is another way to reduce heat loss. Since heat loss increases with surface area, you could determine the best 3D shape to minimize the total surface area for the given volume.

For each of the five basic shapes above, use spreadsheet software to find the dimensions of this shape with the minimum surface area. In order to do this, follow these steps.

a Determine the appropriate equation for the volume of the figure.

b Determine an expression for the height of the figure based on the given volume of $1500\,cm^3$.

c Using substitution, find an equation for the surface area of the figure in terms of another measurement (e.g. radius).

d Use a spreadsheet to calculate the surface area of the figure using various values for the measurement. Find the one that results in the minimum surface area. (Think about how many decimal places you should round to.)

e Use graphing software such as Graphmatica or Desmos or a graphic display calculator (GDC) to graph the equation you found in step 3. Use this to verify the dimensions that give the minimum surface area.

5 Bivariate data

In this unit, you will see how two variables can be used to represent and analyse what it means to be human. However, this same statistical tool can also be instrumental in understanding other ideas, such as democracy and social history.

 Fairness and development

Democracy

What if you could analyse the effects of living in a democracy? What would it tell you? Would your results always be positive? This is an ideal place to apply the study of bivariate data. The strength of the relationship between two variables may help you see why so many around the world are willing to fight for the rights afforded by a democracy.

A democracy is often defined by the equality of all people, the protection of their rights and the representation of citizens in government. The spread of this political ideal over the centuries can be easily studied using statistical tools.

The positive effects of democracy on the economy and personal wellbeing can be easily represented and analysed using statistics.

These same tools can also be used to show that decision making in democracies, especially in response to major issues, can sometimes take a much longer time than other systems of government.

Orientation in space and time

Social histories

What must it have been like to grow up as an African American in the 1950s and 1960s, when racial inequality was being identified and fought? What was life like before, during and after schools were desegregated? What effects did these changes have on individuals and on the society as a whole? Statistics can help you understand these very complex changes and relationships.

Studying the social history of Indigenous Australians can shed light on important aspects of life, such as natural medicines, mental health and how to eat healthily and use resources responsibly. By comparing the results of our current practices to theirs, you will see that we have much to learn from one of the oldest cultures on Earth.

5 Bivariate data
What it means to be human

Related concepts: Models, Quantity

Global context:

In this unit, you will explore the very nature of what it means to be human. As you broaden your study of the global context of **identities and relationships**, you will see that representing and analyzing data can help you to describe the nature of human existence. You will focus on exploring the relationships between many different aspects of the lives of the people on this planet.

Statement of Inquiry

Modeling the relationship between quantities can highlight what it means to be human.

Objectives

- Representing bivariate data using a scatter plot
- Representing data using a line of best fit
- Calculating Pearson's correlation coefficient
- Analyzing data and drawing conclusions

Inquiry questions

F
What is a model?
What is a cause-and-effect relationship?

C
How are relationships modeled?
To what extent can you prove a cause-and-effect relationship?

D
How can the human experience be quantified?
What does it mean to be human?

You should already know how to:

1 Set up a graph with appropriate scales and plot points

Draw a set of axes with appropriate scales to plot the following points.

(−3, 4)   (12, −30)   (4, −17)   (22, 0)

2 Determine the mean, median, mode and range of a set of data

What is the mean, median, mode and range of the following data set?

4   5   2   10   16   8   7   5   2   1

3 Classify data as ordinal, categorical, discrete or continuous

Classify each of the following data types as either ordinal, categorical, discrete or continuous.

a foot size   **b** gender

c temperature   **d** movie rankings

4 Find the equation of a line given two points

Find the equation of the line passing through each pair of points.

a (−1, 3) and (4, −2)   **b** (2, 0) and (4, −6)

c (1, 1) and (−2, 2)

5 Write the equation of a line in standard form, gradient–intercept form and point–gradient form

Write the equation of the line passing through (−2, 4) and (2, −8) in standard form, gradient–intercept form and point–gradient form.

Introducing bivariate data

What does it mean to be human? What makes us different than other species? How can we define and describe the human experience? Answering these questions may be more difficult than you think. However, the very notion that you can ask and reflect on such questions is, in fact, part of what makes you human.

As you mature, you will examine questions such as "Who am I?", "What do I represent?" and "What do I believe in and stand for?". The answers to these questions develop over time as you gain more experience in life and as your connections to friends and family change. In essence, the answers you find to these questions will not only define who you are in your own life but, on a larger scale, they will also help define your place in humanity.

In this unit, you will use data to represent and analyze what it means to be a human being on this planet. It is unlikely that you will answer all of the questions you have, but you will be well equipped to tackle more and more of them as you get older.

Reflect and discuss 1

- What do you think it means to be human? What are some of the elements of the human experience?

- Give two examples of where the human experience can be quantified (represented by a number).

- Give an example of two variables in your life that are related and describe the relationship between them. (For example: "The more I study, the better my grades are.")

Representing bivariate data

Bivariate data refers to data related to two variables. When you studied mean, median and mode, you were analyzing *univariate* data, which means data related to a single variable. Your study of bivariate data will involve looking at how two quantities are related to one another, such as expected lifespan and access to clean drinking water. Understanding the relationship between two variables can help you to understand and even try to improve the human experience.

Scatter plots

Scatter plots are graphical representations that are used in statistics to investigate the relationship between two variables. They are very similar to graphs you have studied and drawn previously, since they are plotted using a horizontal axis (the *x*-axis) and vertical axis (the *y*-axis). The relationship that is shown by a scatter plot is often referred to as the *correlation* between the two variables.

> Scatter plots are also referred to as *scatter diagrams* and *scatter graphs*.

Consider the following data set which shows the heights of 14 females and their shoe sizes. The scatter plot representation is given on the right.

Height (cm)	Shoe size (U.S. sizing)
168	8
183	11
170	8.5
174	9
178	10
180	11.5
160	7
164	7.5
163	7.5
177	9.5
168	8.5
177	10
185	13
186	12

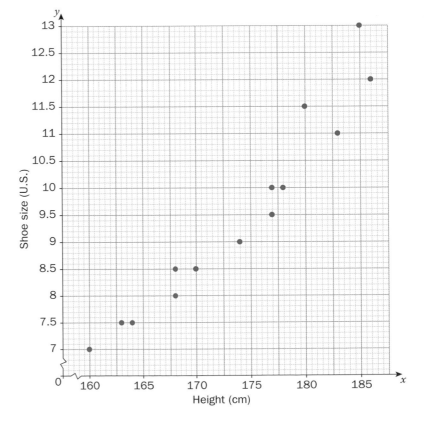

Reflect and discuss 2

- Explain why "scatter plot" is an appropriate name for this type of graph.

- What types of data (categorical, ordinal, discrete and/or continuous) can be represented with a scatter plot? Justify your answer.

- The data that is graphed on the x-axis is called the *independent variable* and the data that is graphed on the y-axis is the *dependent variable*. Why do you think that is?

- The axes are "broken" at the beginning since they do not start at zero. Why is this done?

- Which representation, the table or the scatter plot, is more effective for showing trends in the data set? Explain.

- What are your first impressions about the data? Describe the relationship that you see between height and shoe size among the 14 females.

In order to interpret a scatter plot, you need to look for and describe overall patterns in the data. A scatter plot can be described by its *form (linearity)*, *direction* and by the *strength* of the relationship between the two variables being compared.

Investigation 1 – Describing relationships

Form

The pattern between two variables can be described as having either a *linear* (in a straight line) or *non-linear* form.

1 Look at the arrangement of points on the scatter plots below. Describe the form of each scatter plot as either linear or non-linear. Justify your answers.

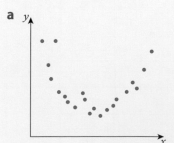
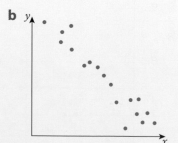
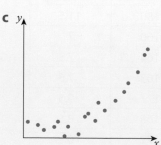

2 Sometimes the data points on a scatter plot do not follow any pattern at all. What would this scatter plot look like? Sketch it.

▶ Continued on next page

3 a For the height versus shoe size data set, describe the relationship as either a linear or a non-linear pattern.

  b Does this make sense given the context of the data set? Explain.

Direction

The direction of the relationship shown on a scatter plot can be described as either *positive* or *negative*.

1 Look at the arrangement of points on the scatter plots below. Describe their direction as either positive or negative. Justify your choices.

a 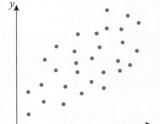   **b**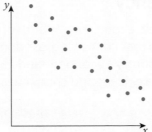

2 What is meant by "direction"?

3 Choose the correct word to complete each statement.

  a When the direction is positive, as one variable increases the other variable **increases/decreases**.

  b When the direction is negative, as one variable increases the other variable **increases/decreases**.

4 a Describe the direction of the height versus shoe size data set. Justify your answer.

  b Does this make sense given the context of the data set? Explain.

Strength

The strength of the relationship between the two variables can be described as *weak*, *moderate* or *strong*.

1 Which of the relationships in the scatter plots below would be considered weak, moderate and strong? Explain.

a 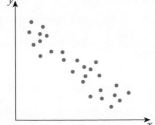   **b**    **c**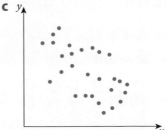

2 Describe how the strength of the relationship is related to how scattered the data points are.

3 What do you think it means when there is a strong relationship between variables?

▶ Continued on next page

4 What do you think it means when there is a weak relationship between variables?

5 For the height versus shoe size data, classify the strength of the relationship between the variables.

Summary

Summarize the information in this investigation in a graphic organizer (e.g. a flow chart), showing the steps you need to go through when analyzing scatter plots.

Activity 1 – Life expectancy

The following data represents the life expectancy at birth of a child in several countries as compared with that country's gross domestic product (GDP). The GDP is a measure of a country's economy, and includes the value of everything produced by all of the citizens and companies in a country.

Country	Gross domestic product (PPP per capita)	Life expectancy (years)
Australia	43 000	82
Bhutan	7 000	68
Canada	43 100	82
Egypt	6 600	73
Japan	37 100	84
Mozambique	1 200	52
Nepal	1 500	75
Oman	29 800	67
Peru	11 100	73
Qatar	102 100	78
United Kingdom	37 300	80
United States	58 000	79

Reflect and discuss 3

In small groups, discuss the following questions.

- What variable will you graph on the horizontal axis? Justify your choice.

- What is the range of each variable? How does this affect how you will set up each of the axes of your scatter plot?

- What scale will you use for each variable? Explain your choice.

▶ Continued on next page

1 Based on your answers to the previous questions, represent the data on a scatter plot.

2 Do the data points line up perfectly? Explain.

3 Would it make sense to join the data points? Explain.

4 What pattern can you see in the data? Describe any general trends that you see.

5 Describe the form, strength and direction of the relationship represented in the scatter plot.

6 Does there seem to be a relationship between a country's GDP and the average life expectancy of its citizens? Justify your answer.

7 Explain what the data seems to imply about what affects the **length** of the human experience.

8 What, if anything, does the data suggest about the **quality** of the human experience? Explain.

When analyzing statistics, you need to ensure that the data you have collected is reliable and valid. The internet is a major source of information, but how do you know that the statistics quoted on a website are accurate? When surfing the web for information, you must think about the trustworthiness of the site.

- Who collected the data? What is their motivation for finding and sharing the data? Is there a bias?
- Is the information on the website current and up to date?
- Does the website list where it sources its information? Does it contain links to other websites that you can check?

ATL1

Pairs

Activity 2 – The importance of reliable data

1 In pairs, find five websites that you think would be trustworthy sources for collecting worldwide data.

2 Check each site for validity and reliability and share your findings with the class.

3 As a class, come up with a list of approved websites you can use to access information.

4 Use the websites you have found to try to answer the following question:

"Is crime related to poverty?"

> To investigate this question, you will need to find the poverty statistics of a country (or city) and the crime rate in that same country (or city).

▶ Continued on next page

Think about these questions.

- How many countries/cities will you need to look at?
- How will you decide on which countries/cities to include?
- Which will be the independent variable and which will be the dependent variable?

5 Represent the data with a scatter plot drawn by hand. What does your scatter plot seem to indicate in response to the question you were investigating?

Practice 1

1 Describe the form, strength and direction of these scatter plots.

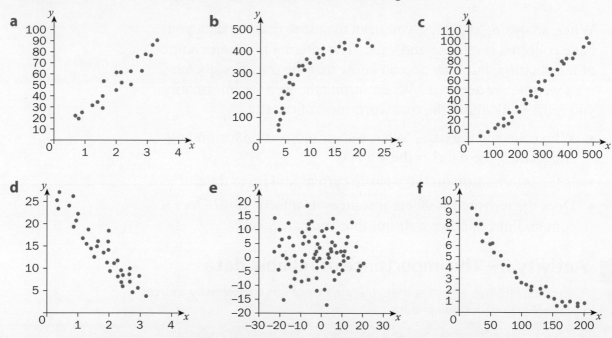

2 Humans are one of the few species that seem to do activities for pure enjoyment. Whereas most animals focus their efforts on survival, humans use some of their time for leisure – simply enjoying things that interest them.

People are prepared to spend a lot of money on their leisure activities. For example, the cost of tickets for a professional basketball game are shown in the following table.

▶ Continued on next page

Number of tickets	Total cost ($)
1	143
2	286
3	429
4	572
5	715
6	858
7	1001
8	1144
9	1287
10	1430
11	1573
12	1716

a Is each data set discrete or continuous? Explain.

b Which is the dependent variable? Justify your answer.

c Which is the independent variable? Justify your answer.

d Represent the data on a scatter plot.

e Do the data points line up perfectly? Would it make sense to join them? Explain.

f Describe the relationship that is represented in the scatter plot (form, strength, direction).

g Represent the data using an equation. Show your working.

3 While birth and death are common to all living things, only humans actually celebrate death. With cultural events like the *Día de los Muertos* (Day of the Dead), humans all over the world have their own way of commemorating those who have died. The *Día de los Muertos* festival in Los Angeles, California, lasts several days from the end of October to the beginning of November. Attendance at the festival over a five-year period is given in the following table.

▶ Continued on next page

Year	Attendance (number of people)
2013	24 882
2014	30 941
2015	33 127
2016	35 081
2017	39 572

a Represent the data using a scatter plot.

b Describe the relationship between the two variables (form, strength, direction).

c Predict the festival attendance for 2018. Justify your answer.

d Suggest reasons why the festival attendance keeps increasing.

4 One of the most notable human traits is the ability to create products, habitats and modes of transportation. However, unlike most other species, we are also capable of destroying natural resources on a large scale. The amount of forest remaining in Indonesia as a function of the country's population is represented in the table below.

Year	Population (millions)	Area of forest (km²)
2007	233	964 870
2008	236.2	958 020
2009	239.3	951 170
2010	242.5	944 320
2011	245.7	937 476
2012	248.9	930 632
2013	252	923 788
2014	255.1	916 944
2015	258.2	910 100

a Represent the data for population and area using a scatter plot.

b What pattern is suggested by the data?

c How do you think the area of forest in Indonesia changed during 2016 and 2017? Explain.

d Describe the relationship that is represented in the scatter plot (form, strength, direction).

e Suggest reasons why the area of forest is decreasing. Is it possible that it is unrelated to the increasing population? Explain.

▶ Continued on next page

5 As humans, we care about our fellow citizens and are willing to give some of our wealth to help those in need. This empathy for others is a key component of what it means to be human and to be part of a community.

The World Giving Index ranks countries according to the percentage of the population who give to charity. The country of Myanmar is often at or near the top of that list. The table below shows Myanmar's World Giving Index score in consecutive years, along with the average temperature in June in Myanmar in those years.

Average temperature in June (°C)	World Giving Index score (%)
25.83	22
25.47	29
25.92	58
25.85	64
25.95	66
26.12	70
26.37	65

a Represent the data using a scatter plot. Be sure to choose suitable scales for your axes.

b Describe the relationship that is represented in the scatter plot (form, strength, direction).

c Do you think the average temperature in June has any impact on the charitable nature of citizens in Myanmar? Is this represented in the graph? Explain.

d What variable could you plot instead of "average temperature in June" that might demonstrate a relationship with the World Giving Index score of Myanmar? Explain.

e If a scatter plot demonstrates a pattern, does that mean that the two variables are necessarily related? Does that mean that one variable necessarily affects the other? Explain.

6 Many species of animals seem to have clearly defined "roles". For example, a beehive has a queen, worker bees, drones, foragers and guards. In contrast, part of the human experience is choosing a job that not only earns money but may also fulfil the need to feel useful and valued.

▶ Continued on next page

Data, like those in the table below, suggest that the number of times a person changes jobs by the age of 30 is changing with each generation.

Generation	Average number of job changes by age 30
Matures	2.1
Baby Boomers	2.4
Gen X'ers	3.7
Millennials	6.4

a Define each variable as either categorical, ordinal, discrete or continuous. Justify your answers.

b Can you represent this data using a scatter plot? Explain.

c What pattern seems to be evident in this data? Explain.

d Do some research and make changes to the independent variable so that you can represent the data using a scatter plot. Draw the scatter plot.

e Compare your scatter plot with three others in your class. Is there more than one correct answer? Explain.

f Describe the relationship that is represented in your scatter plot (form, strength, direction).

g Make a prediction about the average number of job changes for your generation. Justify your prediction.

Line of best fit

In general, the data points on a scatter plot do not all line up perfectly. You will usually need to approximate the relationship in the data with a curve, something called *curve fitting*. This curve (which may be a line) will not go through all of the data points, but it should follow the general pattern of the data so that it represents them as well as possible.

If the curve on your scatter plot is a line, you have drawn the *line of best fit* (often referred to as the *trend line*).

Reflect and discuss 4

- If you had to draw a single line or curve through each of the following scatter plots, what would it look like? Justify your answer.

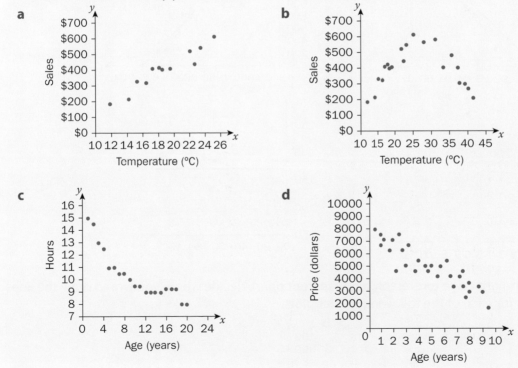

- Which of the above scatter plots suggest a linear relationship?
- For those relationships that are linear, how could you find the equation of the trend line?

Activity 3 – The line of best fit

Perform this activity with a partner.

1 Look at the following scatter plots and lines of best fit. Discuss whether or not you think the line of best fit is well drawn. Justify each of your decisions.

a
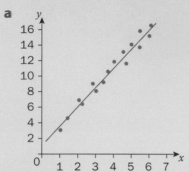

b
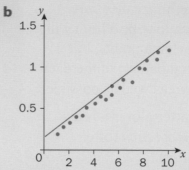

c
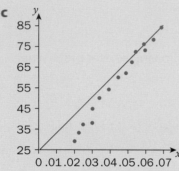

d
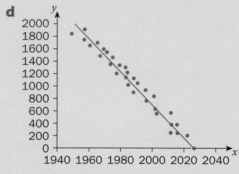

e
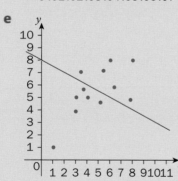

f
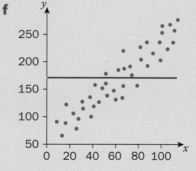

2 Using a straight edge (preferably a transparent ruler), decide where you would draw the line of best fit for each of the following scatter plots.

a
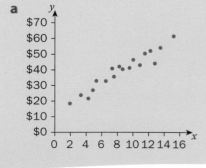

b
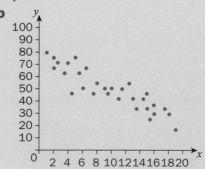

▶ Continued on next page

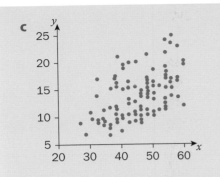

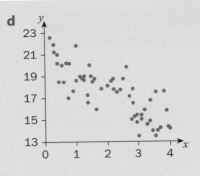

3 Do all lines of best fit have to pass through the origin (0, 0)? Explain.

4 Describe the aspects of drawing the line of best fit that are difficult.

5 Once you have drawn the line of best fit, how can you find its equation?

Drawing a line of best fit can be made easier if you know at least one point on it. If you find the mean of each variable, use that to create an ordered pair, $(\bar{x}, \bar{y})$, and then plot that point. This will be one point that must be on your line of best fit. You still need to try to follow the pattern of the data as you draw the trend line.

> The symbol for the mean or average value of y is $\bar{y}$.

Example 1

Q Humans have a remarkable desire to try to minimize or eliminate the effects of aging. Every living thing ages, but it is humans who have the ability to create products to compensate for this natural process. For example, your vision will slowly deteriorate as you age. The data in the following table show the percentage of the population who wear glasses at specific ages.

▶ Continued on next page

Age (years)	Percentage of population who wear glasses (%)
8	10
14	23
18	28
25	36
35	39
45	55
53	83
60	91
70	92

a Represent the data using a scatter plot.

b Draw the line of best fit and find its equation. Write the equation in gradient–intercept form.

c Use your equation to predict the percentage of the population who might wear glasses at age 30.

 a Begin by calculating the mean point.

$$\bar{x} = \frac{8+14+18+25+35+45+53+60+70}{9} = 36.4$$

> Find the mean of each variable.

$$\bar{y} = \frac{10+23+28+36+39+55+83+91+92}{9} = 50.8$$

The mean point is (36.4, 50.8)

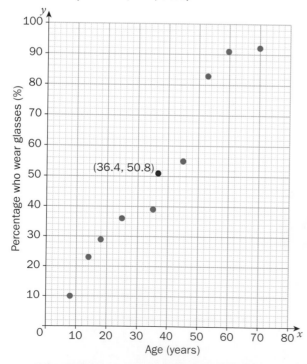

> Set up the scale for each axis. The independent variable (x-axis) is "age", so the x-axis should go from 0 to at least 70. The dependent variable (y-axis) is the percentage of people who wear glasses so the axis should be scaled from 0 to 100.
>
> Plot the points and add the mean point.

▶ Continued on next page

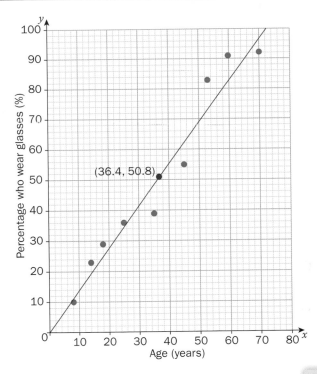

The following callout boxes appear to the right of the graph:

Draw the line of best fit. The line must pass through the mean point and follow the pattern of the data. Try to draw a line with roughly the same number of data points on each side of the line.

The trend line does **not** have to pass through (0, 0), but it does in this case.

b (0, 0) and (50, 70)

$$m = \frac{70-0}{50-0}$$

$$m = \frac{70}{50}$$

$$m = \frac{7}{5}$$

$$y = mx + b$$

$$y = \frac{7}{5}x$$

The equation of the line of best fit is $y = \frac{7}{5}x$

Using two points on the line of best fit, find its equation.

One of the points on the line is the mean point, but you can use any points that are actually on the line. Since the graph appears to go through the points (0, 0) and (50, 70), use these two points. Notice that the points do not have to be the first point and the last one.

Find the gradient of the line.

The y-intercept is (0, 0), so this is the equation in gradient–intercept form.

c $y = \frac{7}{5}x$

$$y = \frac{7}{5}(30)$$

$$y = 42$$

Since you know the age of the person (30), substitute this value for the variable x.

At age 30, roughly 42% of the population will wear glasses.

Representing a set of bivariate data with a linear model and finding its equation is a process called *linear regression*. Therefore, the line of best fit (or trend line) can also be called the *regression line*. If you have the data, you can determine the regression line (line of best fit) using technology as well.

Graphic display calculators (GDCs) or online resources like those at desmos.com and alcula.com allow you to input data and perform linear regression, as was done in Example 1.

Using the Alcula Linear Regression Calculator, you should be able to see the scatter plot and the line of best fit and the program will determine the equation of the regression line.

Activity 4 – How good is your line of best fit?

Being human means not only helping others, but trying to improve the length and quality of human life. While all species adapt to their surroundings, no other species actively searches for cures for illnesses or makes direct attempts to increase lifespan for everyone. Thanks to education and treatment programs, the number of cases of an autoimmune disease (AIDS) in British Columbia, Canada, has decreased dramatically since 2000, as represented below.

Year	Number of AIDS-related deaths
2000	172
2001	151
2002	131
2003	137
2004	130
2005	130
2006	117
2007	110
2008	118
2009	93
2010	72
2011	64

▶ Continued on next page

1 Represent the data using a scatter plot (drawn by hand).

2 On your scatter plot, draw the line of best fit and find its equation. Make sure your line passes through the mean point. Show your working.

ATL2

3 Using a graphic display calculator (GDC) or an online tool, find the equation of the line of best fit.

4 Using a different color, draw the line of best fit you obtained using technology on your hand-drawn scatter plot.

5 How does the line of best fit you drew in step 2 compare with the one you obtained using technology? Explain.

6 How does the scale of your graph affect the placement of the line of best fit? Explain.

Practice 2

1 Find the equation of the line through the following points. Write your answer in gradient–intercept form. Show your working.

  a (−1, 3) and (2, −6) **b** (3, 2) and (−5, 1) **c** (−1, −4) and (6, −5)

  d (12, −9) and (18, 2) **e** (12, 95) and (20, 111) **f** (125, 3) and (61, 10)

2 Find the equation of the line of best fit in each of the scatter plots below.

a

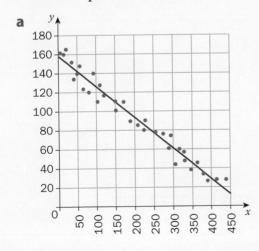

b

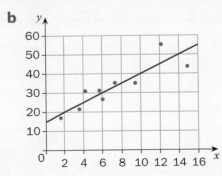

▶ Continued on next page

3 For each of the data sets below, draw the scatter plot and the line of best fit. Find the equation of the line of best fit.

a

x	y
2	13
4	21
7	29
8	35
10	44
12	61
13	70

b

x	y
0	48
1	32
2	35
3	29
4	27
5	19
6	7

c

x	y
23	101
26	98
31	78
33	76
38	67
40	69
44	54
47	51
52	43

d

x	y
12	21
14	23
17	25
18	27
20	28
22	30

4 Draw a scatter plot for data where the line of best fit is easy to determine. Explain your answer.

5 Whether you like it or not, being human also means studying mathematics! While some other species can be taught to answer routine math questions, only humans use mathematical thinking and notation. Getting to this point requires studying mathematics every year. The **total** number of hours a student in Japan has studied mathematics over his lifetime is given for each year from age 6 to 11 years old.

Age (years)	Total time studying mathematics (hours)
6	114
7	269
8	419
9	569
10	729
11	889

a Represent the data using a scatter plot.

b Draw the line of best fit and find its equation. Write the equation in gradient–intercept form.

ATL2 **c** Use technology to find the equation of the regression line.

d How does your hand-drawn line compare with the one you obtained using technology? Explain.

▶ Continued on next page

e Use your equation to predict the total amount of time a 15-year-old student in Japan will have studied mathematics.

f How confident are you in your prediction in step e? Explain.

6 Ours is the only species to be able to communicate ideas to others through a written language. This ability also allows humans to express their creativity, which is another characteristic that is typically human. The recording and exchange of ideas using the written word is an important part of how we pass information on from one generation to the next. The average time spent reading at various ages is given in the table below.

Age (years)	Average time spent reading (hours per day)
17	0.1
22	0.2
30	0.1
40	0.2
50	0.2
60	0.4
70	0.6
75	1.1

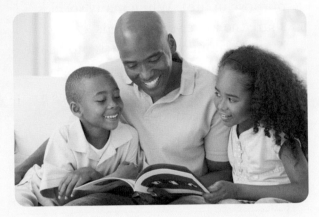

a Represent the data using a scatter plot.

b Draw the line of best fit and find its equation. Write the equation in standard form.

c Use technology to find the equation of the regression line.

d How does your line compare with the one you obtained using technology? Explain.

e Use your equation to predict the age of a person who reads 0.75 hours per day on average.

f How confident are you in your prediction in step e? Explain.

g Describe in words the relationship between a person's age and the average number of hours they spend reading each day.

h Suggest a reason for the relationship you described in step g.

i Does getting older **cause** a person to read more? Explain.

▶ Continued on next page

7 Abraham Maslow, a famous psychologist, created a hierarchy of human needs. He said that humans must fulfil their basic needs, like the need for food and water, before experiencing higher-order needs. Higher-order needs include those related to self-esteem (e.g. the need to feel confident) and to self-actualization (e.g. the desire to be creative or to find purpose in life). While many species share the lower-order needs, it is humans alone who can strive to be the best that they can be.

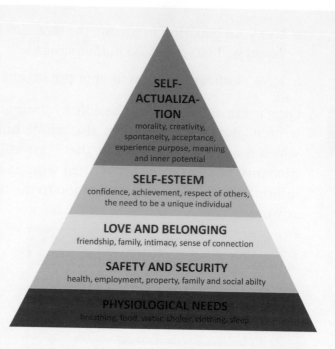

The data in the table below relates to the human need for safety.

Age (years)	Percentage who feel safe walking home at night (%)
20	32.7
30	31.3
42	31.5
57	28.3
70	22.8
75	18.7

a Represent the data using a scatter plot.

b Draw the line of best fit and find its equation. Write the equation in standard form.

c Use technology to find the equation of the regression line.

d How does your line compare with the one you obtained using technology? Explain.

e Use your equation to predict the percentage of students of your age who would feel safe walking home at night.

f Do you agree with your calculation in step e? Explain.

g Describe in words the relationship between a person's age and the likelihood he feels safe walking home at night.

h Does getting older **cause** a person to feel less safe walking alone? Explain.

▶ Continued on next page

Analyzing bivariate data

As you saw previously, you can describe the form, strength and direction of the relationship between two variables. You are focusing on data that follow a linear pattern here, but how do you determine the strength and direction of the relationship? Can they be measured or even calculated?

Describing relationships

Correlation is a measure of the association between two variables. If the relationship appears linear, then it is termed *linear correlation*. The strength and direction of the linear relationship can be represented by a value.

Activity 5 – Describing statistical relationships

Pairs

Perform the following activity with a peer.

1 Below there are seven scatter plots, seven numerical values and seven descriptions of correlation. Match each scatter plot to a value and to a description.

2 Arrange your results in order of correlation, from lowest value to highest.

Weak negative correlation	Strong positive correlation	Perfect positive correlation	Strong negative correlation	Perfect negative correlation	Weak positive correlation	No correlation

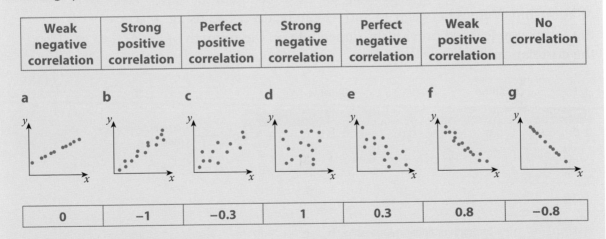

0	−1	−0.3	1	0.3	0.8	−0.8

3 When you have finished, compare your results with another pair of students and summarize your findings.

4 Describe how the value relates to how scattered the data are.

5 For each graph, think of two variables (related to what it means to be human) that might have the given relationship. Explain your choices.

Reflect and discuss 5

Reflect and discuss 5

- How does the value associated with each graph relate to the concept of "gradient" or "slope"? Explain.

- Complete the following sentence. "If the correlation is positive, as one variable increases, the other _____." Explain your answer.

- Complete the following sentence. "If the correlation is negative, as one variable increases, the other _____." Explain your answer.

- If you were to make a prediction based on a scatter plot, which type of scatter plot(s) would give you the most confidence in your prediction? Explain.

The values you sorted in Activity 5 are called *linear correlation coefficients* (*r*) and they indicate how well data can be represented by a line. Data that are perfectly linear have an *r*-value of either +1 or −1, depending on the direction of the relationship. Values between +1 and −1 can be described using the terms you learned previously, depending on the strength of the correlation.

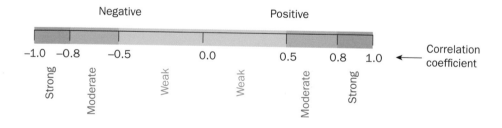

Practice 3

1 Describe each graph as having a positive correlation, a negative correlation or no correlation. For those with a correlation, state whether the correlation is strong, moderate or weak. Justify your choices.

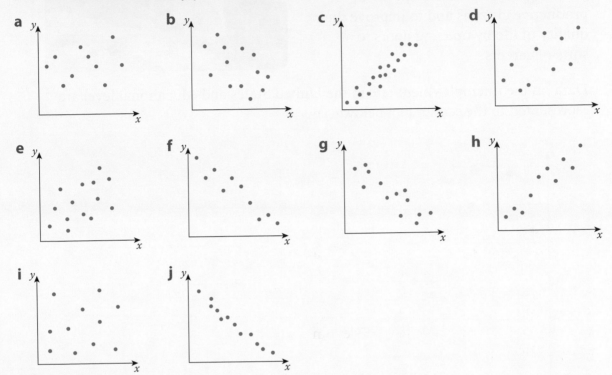

2 Assign a value for the correlation coefficient (r) to each graph in question 1. Justify your answers.

3 What would you expect the correlation between the following pairs of data sets to be? Would you expect a negative correlation, a positive correlation or no correlation? Justify your answer.

   a Level awarded on an assignment versus Number of hours you worked on the assignment

   b Number of cases of the flu versus Number of people immunized against the flu

   c Probability a person has had a romantic relationship versus Age of a person

   d Number of calories consumed versus Number of siblings in family

   e Number of connections in the brain versus Age of an adolescent

   f Number of pets versus Grades in school

4 For the data sets in question 3, classify the relationships that do have a correlation as either strong, moderate or weak. Justify your choices.

▶ Continued on next page

5 All animals learn, but it is only humans that have an organized system of education. While each country adopts its own system, all education systems are designed to produce informed, productive citizens and to improve their quality of life by opening doors to all sorts of careers.

Data on the unemployment rate in the United States and educational level are represented in the scatter plot below.

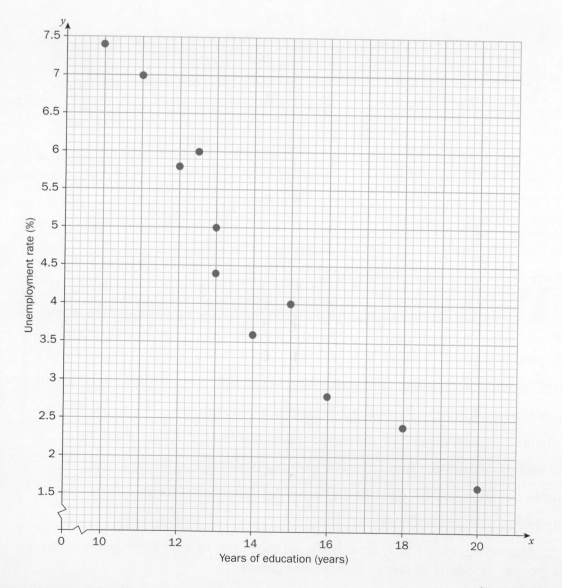

▶ Continued on next page

a Describe the correlation between the two variables as positive/negative and weak/moderate/strong.

b What do you think the value of the correlation coefficient is for this data? Justify your answer.

c What relationship seems to exist between years of education and the unemployment rate? Explain.

6 For each of the following correlation coefficients, write down an example of two variables that might have a correlation with that value. Explain your choices.

0.8   −0.2   1   −0.9   −1   0.5

Calculating the correlation coefficient

Being able to describe a relationship with appropriate mathematical vocabulary is an important skill, but how do you actually calculate the value associated with a correlation? Surely, you are not expected to make a guess every time! One value that can be calculated is the *Pearson correlation coefficient*. It is represented by the symbol r and it measures how well data can be represented by a line.

Did you know...?

The Pearson correlation coefficient is named after Karl Pearson, although his work was based on the work of Francis Galton. Galton pioneered the study of correlation and regression and it was Pearson who continued that work with the calculation of the correlation coefficient. Pearson, a 19th century English mathematician, also wrote about time travel, the fourth dimension and antimatter. Aside from Pearson's impact on the study of statistics, his ideas also influenced the work of Albert Einstein who read Pearson's work.

While there are ways to approximate the value of the correlation coefficient, you can use your GDC or online tools, such as those at www.desmos.com or www.alcula.com, to calculate the correlation coefficient efficiently. The formula for calculating the exact value of r is content that is studied in higher-level mathematics courses.

Activity 6 – Calculating "r"

One characteristic that defines humans is our ability to communicate with one another using language. While other animal species use gestures and sounds to communicate, none have a language like that of humans. The number of words that a child can speak (*expressive vocabulary*) changes as she gets older, as seen in the table below.

Age (years)	Expressive vocabulary (words)
1	20
2	240
3	950
4	1500
5	2200
6	2600
7	3500

1 Enter the data into your GDC or online tool.

2 Draw the scatter plot.

3 Estimate the value of the correlation coefficient. Justify your answer.

4 Use your GDC or online tool to calculate the value of r, the correlation coefficient.

5 How close was your estimate? Are you surprised by the exact value of r? Explain.

Finding the value of r helps you to interpret the potential relationship between two variables. However, what happens if the data include an outlier (a value that simply does not follow the pattern of most of the data)?

Investigation 2 – The effect of an outlier on "r"

criterion B

1 Humans are the only species to use vast amounts of natural resources. No other species extracts more resources from the planet, sometimes putting other species in peril. The population of various countries and their percentage use of the world's resources are represented in the table below.

Country	Population (millions)	Percentage of the world's natural resources used (%)
Thailand	69	2.1
Japan	128	7.3
Germany	82	4.9
Indonesia	261	1.3
Brazil	208	2.9
Mexico	128	2.2
Egypt	96	1.4
China	1403	20.2
France	65	3.5
United Kingdom	66	4.1

a Using technology, draw the scatter plot and find the correlation coefficient. Copy the scatter plot in your notebook.

b Are there any outliers in the data set? Explain how you know.

c Recalculate the correlation coefficient without the outlier included. How has this data point affected the strength and direction of the relationship between the two variables? Explain.

2 Despite our use of natural resources, humans also make an effort to replenish natural resources, for example by planting trees. The country of Myanmar is using drones to increase its ability to replant trees in areas where mangrove trees have been decreasing due to human activity. While some drones map the area, others fly low over the soil and deposit seed pods in desired locations. Since mangrove trees grow in water, scientists have studied the growth rate as it relates to the amount of salt in the water (salinity).

Salinity of water (grams of salt per kg of water)	Growth rate of mangrove (grams/day)
0	0.004
5	0.007
10	0.0036
15	0.0032
20	0.003
23	0.0024

▶ Continued on next page

a Using technology, draw the scatter plot and find the correlation coefficient. Copy the scatter plot in your notebook.

b Are there any outliers in the data set? Explain how you know.

c Recalculate the correlation coefficient without the outlier included. How has this data point affected the strength and direction of the relationship between the two variables? Explain.

3 Describe the effect an outlier can have on the results of linear regression analysis.

Performing linear regression and calculating the correlation coefficient is a useful process in statistics as it allows you to make predictions based on the data that you have.

Activity 7 – Making predictions

Competition is a fact of life for every species on the planet, whether it be for food or habitat or breeding rights. However, humans are unique in organizing competition on a worldwide scale. The Olympics, for example, bring together the best athletes from around the world in order to find out who is the world's fastest human being. The times for the fastest male sprinters in the 100 m race since 1980 are given below.

Runner	Year	Height (cm)	Time in 100 m race (seconds)
Usain Bolt	2012	1.95	9.63
Justin Gatlin	2004	1.85	9.85
Maurice Greene	2000	1.76	9.87
Donovan Bailey	1996	1.85	9.84
Linford Christie	1992	1.88	9.96
Carl Lewis	1988	1.88	9.92
Allan Wells	1980	1.83	10.25

► Continued on next page

1 a Using a GDC or online tool, calculate the correlation coefficient and find the equation of the line of best fit for the relationship between a runner's height and his time in the 100 m race.

  b Describe the relationship as strong, moderate or weak and as positive or negative.

  c Use your equation to predict the winning time of Hasely Crawford, whose height is 1.87 cm. Show your working.

  d How does the correlation coefficient affect the confidence you have in your prediction? Explain.

2 a Using a GDC or online tool, calculate the correlation coefficient and find the equation of the line of best fit for the relationship between the year and the winning time in the 100 m.

  b Describe the relationship as strong, moderate or weak and as positive or negative.

  c Use your equation to predict the winning time of Hasely Crawford, who won in 1976. Show your working.

  d Are you more or less confident in your latest prediction of Crawford's winning time? Explain.

3 Based on your analysis, which is a better predictor of the winning time in the men's 100 m at the Olympics: the height of the runner or the year in which the race was run? Explain.

Practice 4

1 Find the correlation coefficient for each of the following sets of data. Describe the relationship between the variables as positive or negative and as strong, moderate or weak.

ATL2

a

x	y
4	19
7	12
12	24
15	22
11	25
8	11

b

x	y
32	77
25	88
41	72
18	100
27	89
21	95

c

x	y
2.3	2.5
3.5	7.8
11.7	12.4
9.8	11.3
10.2	12.1
12.9	13.7
15.4	13.9
6.6	5.6

d

x	y
12	39
18	33
13	38
22	29
32	19
19	32
25	26

▶ Continued on next page

2 a Interpret the correlation coefficient for data set d in question 1.

b Find the equation of the line of best fit for data set d.

c Verify that each of the points in data set d is on the line you found in step **b**.

d Can the same thing happen in any of the other data sets in question 1? Explain.

e Find the equation of the line of best fit for the other data sets in question 1 using a GDC or online tool.

f For each line you calculated in step **e**, verify whether or not any of the points are on it. Explain your results.

3 An important part of human existence is seeking out relationships with others. While many relationships are friendly, others become so close that the couple may decide to marry. Ceremonies and parties are used worldwide to celebrate and commemorate this important event in the life of a human.

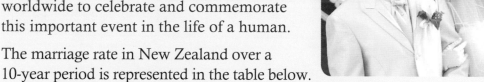

The marriage rate in New Zealand over a 10-year period is represented in the table below.

Year	Marriage rate (number per 100 inhabitants)
2006	14.17
2007	13.98
2008	14.11
2009	13.7
2010	13.04
2011	12.46
2012	12.51
2013	11.44
2014	11.57
2015	11.13
2016	10.95

a Which is the dependent variable and which is the independent variable? Justify your answer.

b Use technology to calculate the correlation coefficient. Round your answer to the nearest hundredth.

▶ Continued on next page

c If you swap the dependent and independent variables, what do you think the correlation coefficient will be? Explain.

d Verify your answer in step **c** by calculating r with the variables reversed.

e Using technology, graph the scatter plot and find the equation of the line of best fit for the original data and for the swapped data. Describe any similarities and differences.

f Explain why it is important to define the dependent variable and the independent variable.

4 a Select two sets of data from the World Bank Data Set (databank.worldbank.org) that relate to what it means to be human.

b Calculate the correlation coefficient between the two variables.

c Classify the strength and direction of the linear relationship.

d Compare your results with those of your peers to determine who has found two variables with the strongest relationship. Suggest one reason why the relationship is so strong.

Correlation and causation

Correlation is used to test and potentially establish a relationship between two variables. *Causation* refers to a change in one variable **causing** a change in the other. **Does a strong correlation imply causation?** Is there a way to prove that a change in one variable causes a change in another? This is one of the most important ideas in the study of statistics.

Activity 8 – Where do babies come from?

Another aspect of being human is that we create legends and myths to explain real-life events. Most cultures have a story of how the Earth was created or your parents may have told you that thunder and lightning was merely the clouds bumping into each other. What did your parents tell you about where babies come from? One common explanation, which originated in Germany many centuries ago, is that storks bring babies. The souls of unborn babies were said to live in marshes and ponds and, since storks

▶ Continued on next page

visit these places often, they could catch the souls and bring them to parents. Can linear regression prove or disprove this often-told tale?

Data on the number of storks and babies born in several European countries is represented in the table below.

Country	Number of babies (thousands)	Number of pairs of nesting storks
Albania	83	100
Denmark	59	9
Hungary	124	5000
Portugal	120	1500
Romania	367	5000
Spain	439	8000
Switzerland	82	150

a Which variable is the independent variable and which is the dependent variable? Explain.

b Represent this data using a scatter plot.

c Find the value of the correlation coefficient using a GDC or online tool.

d What does the correlation coefficient indicate about the relationship between the number of pairs of nesting storks and the country's population? Explain.

e Does this **prove** that storks bring babies? Explain.

f The migration habits of storks actually coincide with periods of increased births in these countries. Does this information change the conclusion that can be drawn from this scatter plot? Explain.

g Give potential reasons for the simultaneous increase in both the human and stork populations.

h Does correlation prove causation? Explain.

Despite two variables appearing to be related, it would require carefully planned experiments to determine whether or not one variable is actually influencing the other. Sometimes, there is a completely different variable that can be used to explain the apparent relationship. This variable is called a *lurking variable*.

Reflect and discuss 6

A study was conducted on the number of cars a person owned and his/her lifespan. The scatter plot is given below, $r = 0.94$.

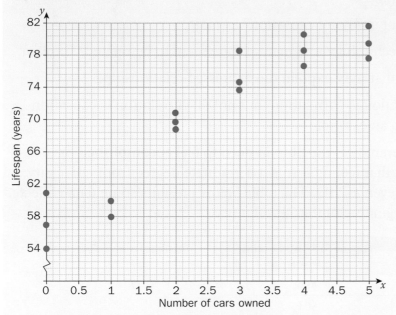

- Describe the relationship between the two variables in terms of form, strength and direction.
- Based on the data, complete the following sentence: "It seems that the _____ cars you own, the _____ you live."
- What lurking variable(s) could be responsible for the simultaneous increase in both the number of cars and the lifespan of the individual? Explain your choice(s).
- If the linear correlation is perfect and all of the data points fall on a line ($r = 1$ or -1), does that indicate that a change in one variable causes a change in the other? Explain.
- Explain what is meant by "correlation does not mean causation"?

Practice 5

1 Which correlation coefficient value(s) indicate that a change in one variable causes a change in the other? Explain.

2 A research study shows a strong correlation between house size and reading ability. Students who live in larger houses tend to score higher on reading ability tests.

   a Isela, who struggles with reading, tries to use the study to get her parents to move to a larger house. What is she hoping will happen? Explain.

   b Does the strong correlation indicate that reading ability is affected by the size of house in which you live? Explain.

   c Suggest a lurking variable that could explain the strong correlation between the two variables.

▶ Continued on next page

3 A university studied students' use of technology and their emotional state. Based on a correlation coefficient of 0.96, researchers concluded that depressed people spend more time sending e-mails and texting.

a Is the researchers' statement implying causation? Explain.

b Why do you think use of technology and mental state seem to be related? Explain.

c Suggest a lurking variable that may explain the expressed relationship.

4 Karl Pearson said, "All causation is correlation, but not all correlation is causation." Explain what he meant using an example.

5 Near the top of Maslow's hierarchy is the human need to feel accomplished or successful. One way to boost that feeling is to get a graduate degree from university, such as a Master's degree or even a PhD. The table below presents some data collected in the United States at the beginning of the 21st century.

Consumption of mozzarella cheese per person (pounds)	Number of PhDs in civil engineering
9.3	480
9.7	501
9.7	540
9.7	552
9.9	547
10.2	622
10.5	655
11	701
10.6	712
10.6	708

a Draw the scatter plot and line of best fit by hand.

b Find the r-value using either a GDC or an online tool.

c Describe the relationship between the two variables (form, direction, strength).

d Use your equation to predict the number of pounds of mozzarella cheese consumed per person in a year when 800 PhDs in civil engineering are awarded.

e Does the strong correlation mean that an increase in mozzarella consumption causes an increase in civil engineering PhDs? Explain.

f Suggest two lurking variables that could explain the simultaneous increase in the two variables.

Pairs

Formative assessment – Does money buy happiness?

Some researchers believe that the impulse to buy and possess things is natural to humans and can be explained by Darwin's theory of evolution. Human beings are materialistic by nature because our early ancestors were "hunter gatherers" who needed to collect resources to survive. Since resources are scarce, humans try to claim as much "stuff" as possible. Money is used to buy these possessions – so does that mean that money can buy happiness?

In pairs, you will compare two statistics to help you to investigate this age-old question, the Happy Planet Index and the Gross National Income (PPP). The PPP shows the wealth of the citizens of a country.

criterion **C**

a What is the Happy Planet Index? What components make up the index? Why do you think those components are included in the index? Do you think any components are missing?

b Select 20 countries to compare in your scatter plot, including the country in which you live. Justify your choice of countries.

c Set up a table of your countries and research the PPP per capita and Happy Planet index value for all 20 countries.

d Which is the independent variable? Why?

e Construct a scatter plot to represent your data. Make sure you have labeled your axes correctly and the variables are on the correct axes.

f Describe your initial observations about the data.

g Draw in a trend line and find its equation (if possible). Discuss what the trend line represents (if possible).

h Determine the correlation coefficient of the data and state the type of correlation. Is a linear trend line an accurate tool to use given these values? Justify why or why not.

i Draw a suitable conclusion based on the value of the correlation coefficient that you obtained and suggest reasons why the points on the scatter plot are not perfectly linear.

> Remember to check the reliability of the source of the information.

Reflect and discuss 7

- Based on your results, does money buy happiness? Justify your response in terms of the indicators you compared and the countries you selected.

- Based on your results, does wealth cause happiness? Explain.

- Describe the aspect(s) of this data set that you find the most interesting.

- When you look at your graph, where does your country lie? Do you feel that this data represents you personally? Explain your answer.

Unit summary

Bivariate data is the name for data related to two variables.

Bivariate data can be represented by a *scatter plot*, also called a *scatter diagram* or *scatter graph*. The relationship that is shown by a scatter plot is often referred to as the correlation between the two variables.

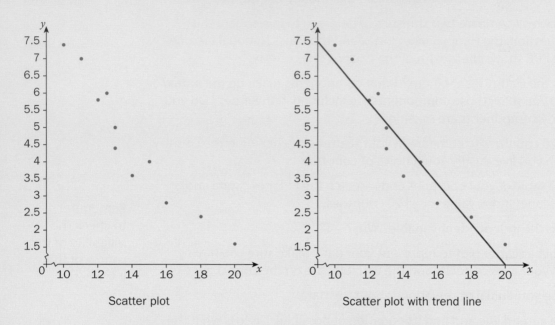

Scatter plot   Scatter plot with trend line

Curve fitting is the name for approximating the relationship between data with a curve (which could be a line). If the curve you use is a line, you have drawn the *line of best fit*, which is also called the re*gression line* or *trend line*. You can approximate the location of the line of best fit by finding the mean of each data set.

Then draw a line of best fit that passes through the mean point and follows the pattern of the data.

The relationship between two variables can be described as either positive or negative. A positive relationship indicates that both variables increase or decrease at the same time. A negative relationship indicates that, when one variable increases, the other decreases.

The *strength* of a linear relationship can be described as weak, moderate or strong. Determining the strength of a linear relationship is easier when you have calculated the *correlation coefficient* (r). The correlation coefficient and the equation of the trend line can both be calculated using a GDC or online tool.

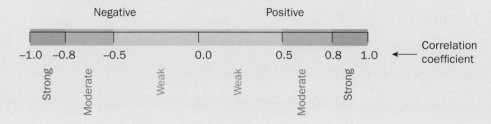

Strength of linear relationship

Correlation does not imply *causation*. Just because a relationship seems to exist between two variables does not mean that a change in one variable **causes** a change in the other.

Unit review

> 🖳 **Launch additional digital resources for this chapter**

Key to Unit review question levels:

Level 1–2   **Level 3–4**   **Level 5–6**   **Level 7–8**

1. For each of the following scatter plots:

   a **describe** the form, direction and strength

   b **suggest** two variables whose relationship might be represented by the scatter plot.

   i

   ii

   iii

   iv

   v

2 a Copy and complete this table.

Variables	Independent variable	Dependent variable	Sketch of scatter plot
1 Number of times you laugh in a day 2 Age			
1 Outside temperature 2 Inside temperature			
1 Numbers of languages learned 2 Number of nationalities in family tree			
1 Son's height 2 Father's height			
1 Reaction time 2 Age			

b In small groups, compare your results and **justify** your answers.

3 Copy and complete these properties of the correlation coefficient:

a The correlation coefficient can take on any values between _____.

b Positive values indicate a _____ association; negative values indicate an _____ association.

c Values closer to 1 and −1 indicate a _____ association; values closer to 0 indicate a _____ association.

d The correlation coefficient only provides an indication of the _____ relationship between two variables. It does not imply _____.

4 Represent each of the following data sets with a scatter plot.

a

x	y
10	12
13	13
15	17
19	19
21	15
24	18
26	20
31	21
25	17

b

x	y
2	44
3	51
6	58
4	31
3	39
5	42
7	27

c

x	y
95	35
87	41
92	40
81	44
83	42
104	22
93	31
90	39

d

x	y
27	77
21	71
22	62
33	51
24	70
28	85
38	56
30	61
20	90

5 The countries with the highest life expectancy for males and females are given in the table below.

Country	Female life expectancy (years)	Male life expectancy (years)
Japan	86.8	80.5
Switzerland	85.3	81.3
Singapore	86.1	80.0
Australia	84.8	80.9
Spain	85.5	80.1
Iceland	84.1	81.2
Italy	84.8	80.5
Israel	84.3	80.6
Sweden	84.0	80.7
France	85.4	79.4
South Korea	85.5	78.8
Canada	84.1	80.2

a Represent the data in a scatter plot. Be sure to include a scale.

b What trend(s) do you see in the data? **Explain** how the scatter plot makes the trend(s) more obvious than the table.

c What aspects of the data are more visible in the table than in the scatter plot? **Explain**.

6 a **Draw** the line of best fit for each of the scatter plots you created in question 4 using the mean point. **Show** your working.

b Find the equation of each line of best fit.

c Use technology to find the equation of the trend line and compare it with your line. How similar are the slope and y-intercept to those you found by hand?

7 A survey was taken in order to compare the outside temperature and people's beverage preferences. Temperature compared with number of bottles of iced tea sold had a correlation coefficient of $r = 0.87$, while temperature compared with number of coffees sold showed the correlation coefficient to be $r = -0.24$. Interpret these results.

8 Unlike most other species, humans live just about everywhere on the planet. We have been able to adapt to a wide range of surroundings, so much so that almost any location is habitable. Why do we choose to live where we do? Does the temperature help people decide where to live? Below is a scatter plot representing data for countries from all around the world in all continents (except Antarctica).

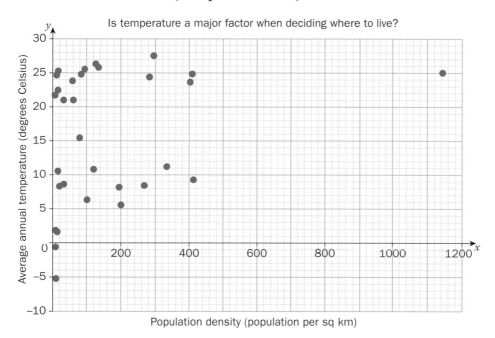

a What does population density tell us?

b What does the average annual temperature of a country tell us? What do you think would be an ideal average annual temperature?

c The data point to the far right is Bangladesh with a population density of 1138 people per square kilometer and an average annual temperature of 25 °C. Would you consider this point an outlier? **Justify** your answer.

d **Describe** the relationship between population density and average annual temperature.

e Is temperature a major factor when deciding where to live? **Suggest** three factors that you think might be more important.

f Are these characteristics good indicators to look at when addressing such a question? What would you alter if you wanted to investigate this question further?

9 One of our most basic human needs is sleep. We spend about a third of our lives doing it! Research suggests that this is because brain cells build connections with other parts of the brain while we are awake and need time to strengthen these connections (and dispose of unimportant ones), which is mostly done while sleeping. As a result, there are many negative health implications when we do not get enough sleep. The table below lists the amount of sleep people, of various ages between 1 and 65, had in a 24-hour period.

Age (years)	Amount of sleep (hours)
2	14
3	13
4	13
4	14
5	12
11	11
12	10
19	9
25	8
27	10
33	8
38	8
42	8
44	7
51	7
60	7
65	6

a Represent the data with a scatter plot. **Describe** the correlation in terms of form, direction and strength.

b Find the equation of the line of best fit using technology.

c What does the slope represent? What does the y-intercept represent?

d Use the equation to **predict** how much sleep a person of your age would have based on this data. How many hours of sleep do you get in a 24-hour period? How far are you from the predicted result?

e According to this data, **predict** how much sleep an 80-year-old would have? Do you think your prediction is accurate?

f How much sleep would you need if you reached the age of 120? Does this answer make sense? **Discuss**.

10 Over half of the world's population lives in cities. We are the only species that urbanize! We see images of big cities and sometimes these cities look highly polluted with dismal air quality. Is that because there are so many people living in such close proximity to each other?

The scatter plot below is a comparison of the pollution index and the population density of cities ranging in size from every continent (except Antarctica). The pollution index is an estimate of the overall pollution in the city (air, water, etc.), with higher numbers indicating more pollution.

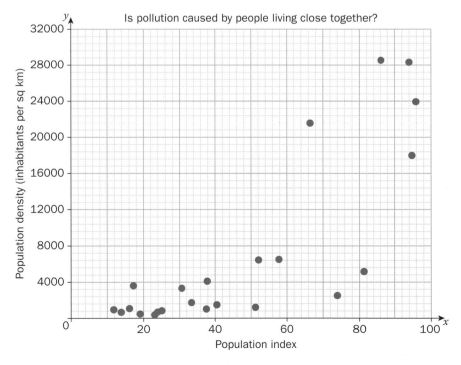

a Was this result what you would have predicted?

b **Describe** the correlation in terms of form, direction and strength.

c Assuming a linear relationship, find the equation of the line.

d Manila is the most densely populated city in the world, with over 71 000 people per square kilometer. What pollution index would you expect this city to have? Does this result make sense?

e In general, you cannot use the scatter plot nor the line of best fit for values that are outside the range of data on the graph. **Explain** why this is.

f Tokyo, Japan, has a pollution index of 46.87. What population density would you expect Tokyo to have? Research the actual population density of Tokyo. Is your prediction close to this? **Discuss** why there is a discrepancy.

g **Suggest** three other factors that would affect the pollution of a city.

ATL1 **11** In this question you will be looking at a factor that may affect human life expectancy.

Decide what factor you would like to analyze and pose it as a question:

"Does _____ affect human life expectancy?"

a Choose a country and research:

- the life expectancy statistics for that country
- the statistics for that country of the other variable that you chose.

b How many countries will you need to look at?

c How will you decide which countries to include?

d Which variable will be the independent and which will be the dependent variable?

e Collect all the data that you need and **draw** a scatter plot. Your teacher will indicate whether you will use technology or create the graph by hand.

f Determine the correlation coefficient. **Use** your results to answer the question you posed at the beginning.

Summative assessment

What does it mean to be human?

You have looked at a wide range of ways to answer this question already, from a scientific perspective to a more ideological approach.

In this task, you will investigate the relationship between two variables that you think would be appropriate to explore this question. You will create a one-page infographic using an online tool such as Infogram or Piktochart that contains the requirements outlined below.

a   Choose two variables whose relationship you would like to investigate. Explain why you have chosen these two variables and how the relationship between them will help you to address the question "What does it mean to be human?"

> The difficulty with such a task lies not only in deciding what variables to compare, but also where to find reliable data to ensure you can analyze the question appropriately. Earlier in this unit, you used a set of guidelines to help you determine some reliable websites. Use the same criteria when finding data for this assessment.

b   How much data will you collect? Who/what will be included? Discuss why you have decided on this amount of data.

c   Where did you source your data? Make sure you include the URLs and names of any websites you used. Justify the reliability and validity of your sources.

d Which is the independent variable? Explain.

e Construct a scatter plot to represent your data by hand or using technology. Make sure you have labeled your axes correctly and that the variables are on the correct axes.

f What are your initial observations about the data?

g Draw in a trend line.

h Determine the equation of the line.

i What do the gradient and y-intercept represent in your equation?

j Determine the correlation coefficient of the data and describe the relationship between the variables.

k Is a linear trend line an accurate tool to use? Justify your answer.

l Summarize your findings by answering the question "What does it mean to be human?"

Reflect and discuss

- How difficult was it to find reliable data?
- What did you learn from this experience that will help you with future research?
- What do you think sets us (humans) apart from the rest of the animal kingdom?

6 Geometric transformations

One way to express beliefs and values is through the application of transformation of various shapes, as you will see in this unit. But in what other contexts could this knowledge be valuable?

Identities and relationships

Physical development

Have you ever wondered what you look like at a molecular level? How do a body and a mind develop over time? What is in our genetic code?

DNA is a long molecule that is made up of smaller pieces, or nucleotides, that are strung together. Each nucleotide is comprised of smaller molecules, some of which carry your genetic information. These molecules are repeated throughout your body and are excellent examples of translations, rotations and reflections, all of which can be studied mathematically.

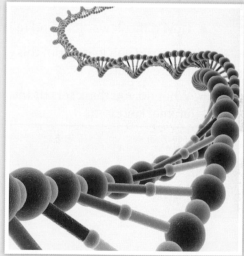

Even the colds you get and the viruses you catch can be analysed through the lens of geometric transformations. The way in which drugs attack these organisms can be represented and studied through transformational geometry. Not even bacteria are a match for mathematics!

Scientific and technical innovation

Digital life

Your digital life includes all of the activities that you perform online, from playing games to uploading photos. Did you know that many of your on-screen activities utilize mathematics? Analysing your digital life can be a great place to look at the applications of geometric transformations.

Video games such as Tetris require you to rotate and translate pieces in order make them fit together. Understanding the results of these transformations can not only make the game easier to play but it may just earn you the high score!

Even the photos you edit and then post online involve the application of transformational geometry. Rotating a photo, reflecting it or applying small translations can produce a whole new look and style.

6 Geometric transformations
Expressing beliefs and values

KEY CONCEPT: FORM

Related concepts: Patterns, Space

Global context:

In this unit, you will see how form and shape are used to reflect and represent different cultures. As part of the global context of **personal and cultural expression**, you will explore how creativity can be enhanced through an understanding of geometric principles and then be used to express a person or organization's beliefs and values.

Statement of Inquiry

An understanding of patterns created by forms in space can enhance creativity and help express beliefs and values.

Objectives

- Transforming a figure by rotation, reflection, translation and dilation
- Analyzing the defining features necessary to produce different types of tessellations
- Applying mathematical strategies to solve problems involving geometric transformations, similarity and congruency
- Creating a tessellation

Inquiry questions

F What is a transformation?
What defines a pattern?

C How are patterns created by different forms in space?

D What enhances creativity?
How do we express culture, beliefs and values?

ATL1 Self-management: Reflection skills

Consider personal learning strategies

- What can I do to become a more efficient and effective learner?
- How can I become more flexible in my choice of learning strategies?
- What factors are important for helping me learn well?

ATL2 Communication: Communication skills

Use and interpret a range of discipline-specific terms and symbols

You should already know how to:

1 Plot points on a coordinate grid

Plot the following points on a coordinate grid.

$(-3, 2)$ $(-5, -5)$ $(0, 7)$

2 Draw an angle

Draw the following angles

a $90°$ b $180°$ c $270°$

3 Draw lines

Draw the following lines on a coordinate grid and label them.

$x = -2$ $y = 3$ $y = x$ $y = -x$

4 Find measures of interior angles of regular polygons

Find the measure of the interior angles of each of the following regular polygons.

a pentagon   b hexagon
c heptagon   d decagon

Introducing geometric transformations

Transformations are all around you. Any time a shape is moved, it is a transformation. You have no doubt seen many transformations before, but perhaps you referred to them using terms such as "flips", "slides" and "turns".

The hands on a clock face demonstrate rotation.

A beautiful landscape over still water shows a reflection.

Translations can be seen in the notes on sheet music.

When transformations are repeated, a pattern is created that can be incredibly beautiful. At the same time, patterns formed by transformations can help represent and express the beliefs and values of a person or a group of people. Whether the patterns are in the Alhambra in Spain or India's Taj Mahal or anywhere else, the common link behind each of these patterns is the mathematics we use to describe them.

Reflect and discuss 1

- List three examples of patterns made up of shapes that you have seen in your daily life.

- Do these patterns involve reflections, rotations, translations or a combination of these? Explain.

- Do you think these patterns reflect the culture of where you live? What does each pattern make you think of, for example, a particular place, idea or feeling?

Tessellations

A *tessellation* is a pattern of geometrical shapes that has **no** overlaps and **no** gaps. The shapes in a tessellation are also referred to as *tiles*. They can be turned (rotated) or flipped (reflected) if necessary to make the pattern. These are all examples of tessellations.

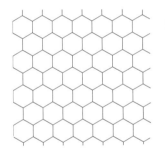  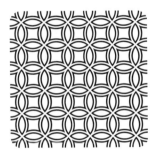 

Reflect and discuss 2

- Explain how each of the examples shown here fits the definition of a tessellation.

- Tessellations involve a single shape being repeated over and over. What is the shape that has been tessellated in each of the above examples? Explain.

- Is there more than one possibility for the tessellating shape in each example? Explain.

Activity 1 – Which shapes tessellate?

1 Look at the shapes below. (Your teacher may give you cut-outs.) Which shapes could you use to tile a floor with no gaps or overlaps? You may turn or flip the shape if necessary. Justify your choices.

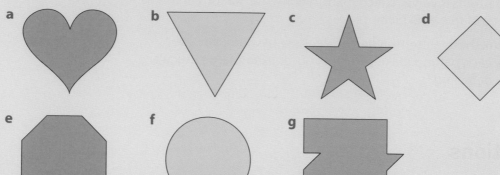

a   b   c   d

e   f   g

Search for "Tessellation Creator" on the NCTM website: illuminations.nctm.org. Use it to investigate the kinds of shapes that tessellate and those that do not. You can drag your choice of shapes onto the screen and position them to make patterns. Experiment with the tools for duplicating, rotating and coloring the shapes.

The Alhambra palace, in Granada, Spain, contains some of the most famous examples of tessellations. The tiles in the Alhambra were laid out by both the Moors (Muslims from Spain and northern Africa) and by Christian artisans who were inspired by the Moors' style in the 14th century.

Pairs

2 With a partner, discuss the following questions.

   a What original shape has been tessellated to create the pattern shown in the photograph?

   b How was the original shape altered so that it would tessellate? (e.g. flipped, turned)

ATL2

3 Research the Alhambra and two other examples of Islamic architecture. Copy and paste examples of the art found in these buildings into a Word document and add a commentary, explaining why each piece represents a tessellation.

Reflect and discuss 3

- Why do you think geometry was used so much in this style of art?
- Where else have you seen geometric shapes being used to create tessellations? Give specific examples.

Congruence transformations

Tessellations are patterns of shapes. Sometimes the shapes have to be modified so that they can tessellate. The ways of changing a shape or form are called *transformations*. Examples of transformations include moving the shape to a new location or enlarging the shape. When a shape is moved to a new location without changing its size, it undergoes a *congruence transformation*. Congruence transformations are the focus of this first section.

Translations

A geometric *translation* moves a shape left or right and/or up or down. The translated shapes remain exactly the same size as the original, so the shapes are *congruent* to each other. They have just been shifted in one or more directions.

Activity 2 – Maze

1 Describe the path you need to take to get from the left side of the maze to the right side of the maze. (e.g. "Move 2 cm up, then move 3 cm right, …") Use a ruler to measure the length of each movement (translation). You can only move up, down, left and right.

2 Now design a maze of your own, no bigger than 20 cm by 20 cm and list the translations needed to get through the maze on a separate piece of paper. Remember, you can only move up, down, left and right.

3 Your teacher will collect the mazes and redistribute them among the class. You will try to find the path through the maze of one of your peers, listing all translations (movements) as you go.

4 When you have finished, find the creator of the maze and compare your path with the answer they wrote down.

Pairs

When describing how a shape has been transformed, the original shape is called the *preimage* and the vertices are often labeled using uppercase letters (e.g. *ABCD*). The translated shape is referred to as the *image* and the vertices are labelled using uppercase letters with a "prime" next to each (e.g. *A′B′C′D′*, pronounced "*A*-prime, *B*-prime, *C*-prime, *D*-prime").

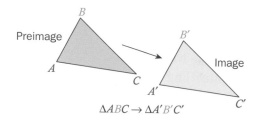

$$\triangle ABC \rightarrow \triangle A'B'C'$$

Example 1

Q Specify the translation of this shape that has occurred. The preimage is the blue figure in the upper left-hand corner.

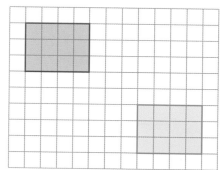

A 7 units right

Select any point on the preimage – in this case, the rectangle in the top left – and count how many units you have to go right or left.

5 units down

Count how many units you have to go up or down.

Translation: 7 right and 5 down

Record the translation using proper notation.

Q Translate the shape on the grid below 8 units to the right and 3 units up.

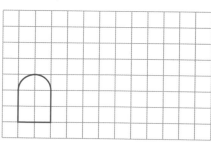

▶ Continued on next page

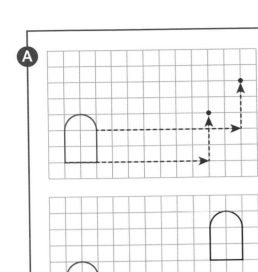

A

Select any point on the preimage and count the number of units required for the translation in the direction given. In this case, go 3 units up and 8 units to the right.

Repeat for all the other points.
Draw the exact image of the original shape.

It is also possible to translate shapes using the Cartesian plane. The structure of the Cartesian plane allows each point on the preimage to be plotted in space and then its image plotted after a given translation. Specialized notation is used to describe the translation.

ATL2

Activity 3 – Translations on the Cartesian plane

1 Copy this table. Translate each of the following preimage points as indicated and enter the new coordinates in the table.

Preimage point	Image 3 units left	Image 7 units up	Image 2 units right, 1 unit down	Image (Create your own)
(3, 9)				
(−2, 7)				
(8, −11)				
(−5, −4)				
(0, 0)				
(0, 10)				
(−12, 0)				

2 What happened to the value of the *x*-coordinate of each point under the first translation? Explain.

▶ Continued on next page

3 What happened to the value of the y-coordinate of each point under the second translation? Explain.

4 a What happened to the value of the x and the y-coordinate of each point under the third translation? Explain.

   b How would you represent this translation? Create your own notation to show what happened to each point.

5 The correct notation for representing translations on the Cartesian plane is shown in this example:

$$(x, y) \rightarrow (x+6, y-4)$$

Describe what this translation does to all of the points in a preimage.

6 Describe the three translations in the table using the correct notation.

7 Create a translation for the final column of your table. Use correct notation to describe your translation. Use your translation to translate each of the preimage points in the table; record the location of each image point.

Practice 1

1 For each shape, specify the translation that has occurred. The preimage is blue and the image is green.

a

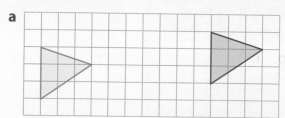

b

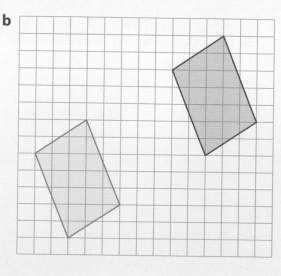

▶ Continued on next page

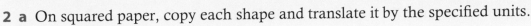

2 a On squared paper, copy each shape and translate it by the specified units.

i

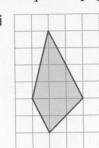

5 units down, 2 units left

ii

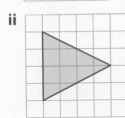

6 units up, 3 units right

iii

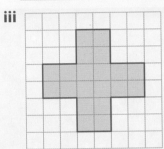

1 unit up, 7 units left, 4 units down, 9 units right

b Could the instructions for translation **iii** be simplified so that the translation can be completed in fewer steps? Explain.

c Write the three translations using correct notation.

3 The logo of the company Auto Union AG is made up of four rings, as shown here. The rings represent the four original car companies that, together, make up Auto Union AG: Audi, DKW, Wanderer and Horch. Out of respect for each of the original companies' contributions, they are all represented in the symbol.

Describe how to produce each of the other three rings if they are all translations of the ring on the right. Use correct notation for each translation.

▶ Continued on next page

4 Plot the shape with vertices (1, 8), (−3, −5), (−4, 7) and (−6, −2).
Then plot the image of this shape after each of these translations.

$(x, y) \rightarrow (x − 3, y)$

$(x, y) \rightarrow (x, y + 5)$

$(x, y) \rightarrow (x + 6, y + 1)$

$(x, y) \rightarrow (x − 2, y − 4)$

5 Given the translation $(x, y) \rightarrow (x + 7, y − 3)$, find the image of each of these preimage points.

   a (5, 9)   **b** (−2, 0)   **c** (0, −8)

6 Given the translation $(x, y) \rightarrow (x − 4, y + 1)$, find the preimage of the following image points.

   a (−6, 15)   **b** (13, −11)   **c** (0, 0)

7 Pac-Man was a popular video game in the 1980s. It was the first time a video game creator made a deliberate attempt to attract more girls to video games, which was very innovative at the time. It later went on to become a social phenomenon, with its own merchandise, an animated television series and even a pop song. Pac-Man's legacy can still be seen in many of today's video games.

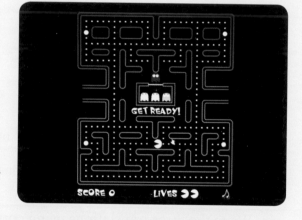

Pac-Man can only move left, right, up and down. List the translations that Pac-Man has to make to eat all four large dots using the shortest path possible (avoiding the ghosts). Count each dot as one unit.

ATL2 **8** This is the Gherkin building in London, UK. In order to be energy efficient, the building was designed to use half as much energy as most large buildings. Many of its features (such as its curved shape and the gaps in between floors creating a natural ventilation system) express the concept that architecture can be both functional and creative.

   a If the vertical height of each diamond-shaped section is approximately 20 m, describe the translation of this shape on the building.

   b Is the pattern on the building an example of a tessellation? Explain.

▶ Continued on next page

9 a This wall has been built using standard bricks, 230 mm wide and 76 mm high. What translations of the brick shape have taken place?

b Is this an example of a tessellation? Explain.

10 This is a Peruvian-style rug. In Peru, rugs and other textiles were more than just artistic products. They were often used to communicate personal histories and relationships between people and their gods and the world around them. Peruvian identity can be found expressed in many textiles from the region.

a Identify the original shape(s) and describe the transformations that have occurred.

b Is this an example of a tessellation? Explain.

Indicate translations left, right, up and down.

Rotations

Geometric *rotations* involve turning an object around a point, called the *center of rotation*. When you rotate a shape, you need to define the center of rotation, the number of degrees you wish to rotate the shape and the direction in which the rotation occurs. You can rotate an object by however many degrees you wish. Positive rotations go in a clockwise direction and negative rotations go in a counterclockwise direction. As with a translation, the image is exactly the same size as the preimage, so the two shapes are congruent.

Rotations are most often described as:

clockwise (CW)   or   counterclockwise (CCW).

Drawing the image after a rotation can be tricky. After rotating on a coordinate grid, make sure each vertex and side of the image are the same distance from the center point as on the preimage. This will ensure that the image is congruent to the preimage.

Investigation 1 – Rotations on the Cartesian plane

criterion **B**

Rotations can also be done on a Cartesian plane. Although a shape can be rotated around any point, at this stage you will always use the origin as the center of rotation. You will focus on rotations of 90, 180 and 270 degrees.

1 Consider the Cartesian plane. What is the measure of the angle formed by the x- and y-axes? Justify your answer.

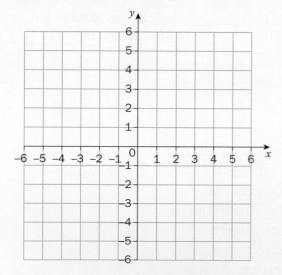

2 Moving in a clockwise direction from true north (the positive y-axis), find the size of the angle turned through to reach each of the following.

  a East (the positive x-axis)

  b South (the negative y-axis)

  c West (the negative x-axis)

▶ Continued on next page

Go to ct.excelwa.org, select the APPS menu, and choose Polygon Transformer. You can use this app to perform all of the geometric transformations in step 3 using any given shape or point. To plot a single point, enter an ordered pair and click "Draw". To draw a line or a shape, input two or more ordered pairs which are then plotted and connected together with straight lines. You can then rotate the point or shape around the origin by 90°, 180° or 270° in either a clockwise (CW) or a counterclockwise (CCW) direction.

3 Copy and complete this table. Rotate each of the following points around the origin through the indicated number of degrees.

Point	90° CW	90° CCW	180° CW	180° CCW	270° CW	270° CCW
(3, 9)						
(−2, 7)						
(8, −11)						
(−5, −4)						
(0, 0)						
(0, 10)						
(−12, 0)						

If you are doing the rotations on paper, a good procedure to help you to visualize the rotation is to graph the original point, then physically "rotate" your graph paper through the given rotation around the origin. The new location of your point represents the coordinates of the image.

4 Copy this table. Describe in words the effect on a point of each of the following **clockwise** rotations about the origin. Then, generalize the results using the correct notation.

Degree of rotation	Result of rotation	Notation
90° CW		$(x, y) \rightarrow (\quad , \quad)$
180° CW		$(x, y) \rightarrow (\quad , \quad)$
270° CW		$(x, y) \rightarrow (\quad , \quad)$

5 Verify your results for two points of your own choosing.

6 Justify why each rule works.

7 While technology makes drawing easier, are there any advantages to drawing rotations by hand? How could this skill make you a more effective learner?

Reflect and discuss 4

- In a rotation, do all points on the preimage move? Do they all move the same distance? Explain.

- Name two different rotations that produce the same image. Explain why this happens.

- When someone completely changes their opinion or actions, it is common to say, "They've gone a complete 360 degrees." Explain why this makes little sense. What would be a more logical way of describing the change?

Practice 2

1 Islamic art is famous for its use of geometric patterns as opposed to depicting living things. This helps express the Muslim idea that artists should not attempt to copy God by creating life or images of life. This picture shows tiles on a wall in the Alcazar of Seville, Spain, a palace constructed by the Moors.

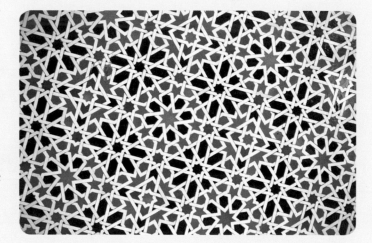

a Does this represent a tessellation? If so, what shape has been tessellated?

b Identify and describe two examples of translations in this pattern.

c Identify and describe two examples of rotations in this pattern. Indicate the centers of rotation and angles of rotation.

▶ Continued on next page

2 Use squared paper to graph the image of each figure using the transformation given.

a Rotation of 180° about the origin

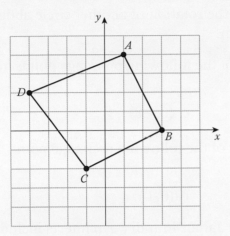

b Rotation of 180° about the origin

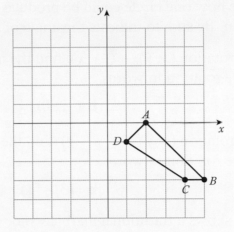

c Rotation of 90° counterclockwise about the origin

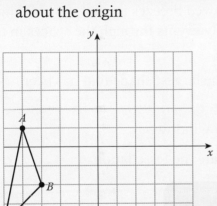

d Rotation of 90° clockwise about the origin

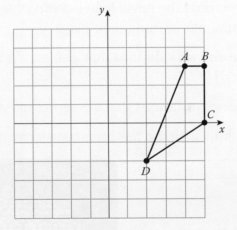

ATL2

3 The Olympic rings express the ideals behind the Olympic Games. Each of the five rings represents a continent, with a different color for each. The rings are interlocked to represent how the Olympic Games unite athletes and people from all over the world.

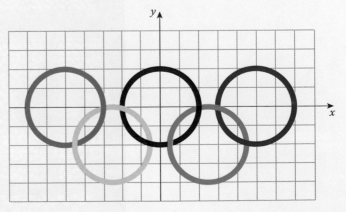

a Describe how each of the other circles could have been translated from the blue circle. Use correct notation.

▶ Continued on next page

b Describe how one pair of circles could have been translated to produce another pair of circles. Use correct mathematical notation.

c Describe how one circle could be produced from the rotation of another circle about the origin. Be sure to indicate the angle of rotation.

4 Given the rotation $(x, y) \rightarrow (-y, x)$, find the image of the following preimage points and describe the rotation in words.

   a (5, 9)   **b** (−2, 0)   **c** (0, −8)

5 Given the rotation $(x, y) \rightarrow (-x, -y)$, find the image of the following preimage points and describe the rotation in words.

   a (−6, 15)   **b** (13, −11)   **c** (0, 0)

6 The photo shows a traditional marble window from Jaipur in India. Architects included these ornate windows because they incorporated two important values. With soaring temperatures in the summer, the open windows allowed breezes to cool the building. At the same time, the royal ladies could observe daily events through the spaces in the pattern while maintaining their privacy.

   a Describe any tessellations that you see. What is the largest shape that has been tessellated?

   b Describe any rotations that are represented in this window.

   c Explain how a shape has been transformed to create the tessellation.

▶ Continued on next page

7 This building is in Ravensbourne University London, UK – a university for digital media and design. The building, including its tiling, was created to express current ideas in design and creation. It was also designed to encourage collaboration among the different disciplines studied at Ravensbourne.

a Describe any transformations that you can see in the tile design.

b Identify the original shape(s) and explain how the shapes have been tessellated.

Activity 4 – Turtle Academy

Back in the 1980s, Logo was an educational computer language that students used to give directions to a small onscreen turtle icon which followed the instructions to produce line graphics. The aim was for students to understand, predict and explain the turtle's movements through the use of computer commands. The name Logo was derived from the Greek word "logos" which means "word" or "reason". It was chosen to express the idea that logic-based Logo was different from other programming languages which were number based.

Well, what was old is now new again and the Turtle is back!

 Head over to turtleacademy.com. In the top bar, select the "Lessons" tab to learn how to control the turtle. Go through as many of the tutorials as you need in order to make the turtle turn, go back, move forward, etc. to make a shape.

Here is a basic example giving a square that is 50 units long on each side.

forward 50 | right 90 | forward 50 | right 90

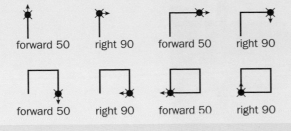

forward 50 | right 90 | forward 50 | right 90

▶ Continued on next page

1 Once you have gone through the lessons and feel comfortable with the commands, you can then make a shape that you can translate. Feel free to be creative.

Here is an example to get your creative juices flowing.

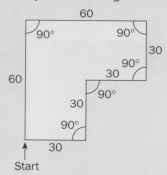

2 When your shape is completed, translate your shape 100 units to the right.

3 Translate your shape 250 units down.

4 Can you rotate your shape 180 degrees clockwise? Explain.

5 Can you rotate your shape 90 degrees counterclockwise? Explain.

6 Now that you've had an opportunity to play around in Turtle World, can you think of where this software and these skills can be applied today?

Reflections

Reflections are all around you; when you look in the mirror you see your own reflection.

Take a look at this photo of a human landscape reflected in water, and at the reflection of a natural landscape on page 258.

Reflect and discuss 5

- In the photos, which is the image and which is the preimage? Explain.

- How does the size of the preimage compare with the size of the image? Explain.

- A rotation requires a point of rotation, an angle and a direction of rotation. What is necessary to define a reflection?

A *reflection* is when a point or shape is flipped across a *line of reflection*, also known as a *mirror line*. In a reflection, each point in the image is the same distance from the line of reflection as the corresponding point in the original preimage.

Example 2

(Q) Draw the line of reflection.

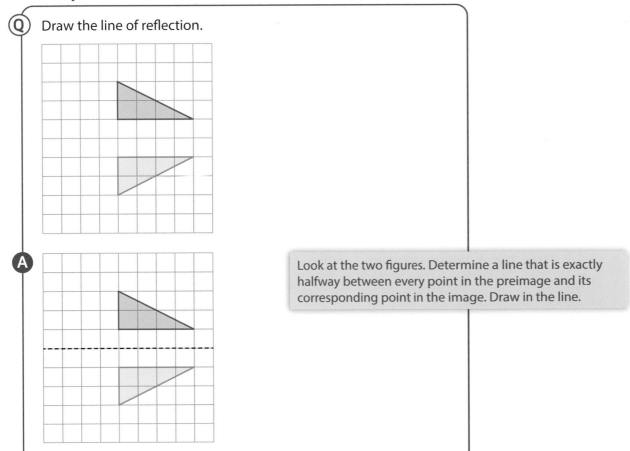

(A)

Look at the two figures. Determine a line that is exactly halfway between every point in the preimage and its corresponding point in the image. Draw in the line.

▶ Continued on next page

Q Draw the image after being reflected in the given mirror line.

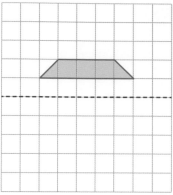

A

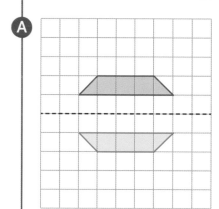

First, measure the distance from the vertices of the preimage to the line of reflection.

Then, mark the vertices of the image on the opposite side of the line of reflection, the same distance away. The image will be identical in size and shape, simply flipped.

Sometimes reflections seem easier if you turn your paper so the line of reflection is vertical.

Reflect and discuss 6

ATL2

- Why is a reflection a congruence transformation? Explain.

- Where is the line of reflection in relation to both the image and the preimage? Explain.

- Where do you see reflections in the art and architecture of your school? What do you think is the purpose of these reflections?

Practice 3

1 Copy these reflections on squared paper and draw the line of reflection on each.

a

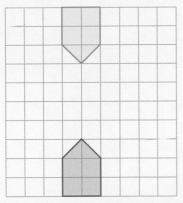

b

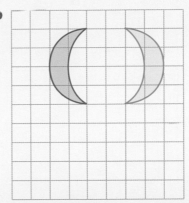

2 This image shows the Under Armour logo. It has no cultural significance, as it simply represents the initial letters of the company name in an eye-catching way: a U with a stylized A underneath. Copy or trace the logo.

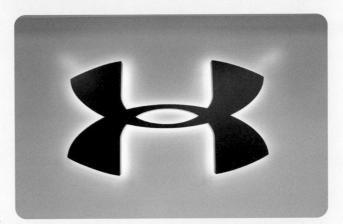

a Explain how the logo can be created using a reflection. Indicate where the line of reflection is.

b Explain how the logo can be created using a rotation. Indicate the center of rotation and the angle of rotation.

3 A coat of arms is a symbol that represents a person, family or country. Coats of arms are often placed on shields or flags. The Serbian coat of arms is shown on the shield in this photo. Two of the main symbols in it, the double-headed eagle and the Serbian cross in the middle, represent the national identity of Serbian people throughout history. The four stylized Cs around the cross refer to a popular motto that reminds Serbian people of their ability to save their heritage and nation if they all work together.

a Describe the reflection that was used to create the coat of arms. Indicate the location of the line of reflection.

b Describe two examples of translations that were used to create the coat of arms.

▶ Continued on next page

4 Copy the shapes and mirror lines onto squared paper. Draw the reflected image of each shape.

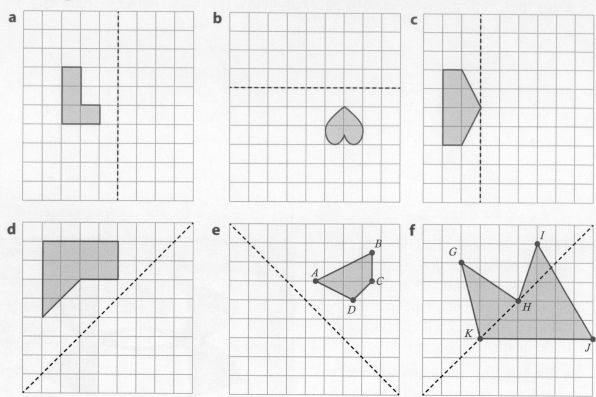

a

b

c

d

e

f

 Investigation 2 – Reflections on the Cartesian plane

criteria
B, C

The easiest way to perform a reflection is simply to draw in the reflection line. Then draw each point exactly the same distance away from the line, but on the opposite side, so that the mirror image is identical to the original preimage.

You can use the Polygon Transformer app that you used in Investigation 1 to perform most of the geometric transformations in this investigation. Input one ordered pair to plot a point, or several ordered pairs to plot a shape. Then click "Draw" to plot your point or draw your shape. You can then reflect the shape or point in the x-axis, the y-axis or any other vertical or horizontal line.

You won't be able to reflect in the lines $y = x$ or $y = -x$ using the app, so do these two reflections on paper.

Go to ct.excelwa.org, select the APPS menu, and choose Polygon Transformer.

▶ Continued on next page

1 Reflect each of the given points in the following reflection lines.

Point	x-axis	y-axis	x = 1	y = 3	y = x	y = −x
(3, 9)						
(−2, 7)						
(8, −11)						
(−5, −4)						
(0, 0)						
(0, 10)						
(−12, 0)						

2 Copy and complete this table. Describe in words the effect of each of these reflections on a point. Then generalize the result using correct notation.

Reflection line	Result of reflection	Notation
x-axis		$(x, y) \rightarrow (\quad , \quad)$
y-axis		$(x, y) \rightarrow (\quad , \quad)$
$y = x$		$(x, y) \rightarrow (\quad , \quad)$
$y = -x$		$(x, y) \rightarrow (\quad , \quad)$

3 Verify each rule for two more points of your choice.

4 Justify your rule.

> If you are doing the reflections on paper, a good method to help you to visualize the reflection is to graph the original point, then physically fold your graph paper along the reflection line. The new location of your point represents the coordinates of the image.

Reflect and discuss 7

Make a prediction about the following combined reflections, and then try an example to test whether your predictions are correct:

- What single transformation produces the same result as consecutive reflections over two parallel lines?

- Would the same result be achieved over any even number of parallel lines? Explain.

- What single transformation produces the same result as consecutive reflections over three parallel lines?

- Would the same result be achieved over any odd number of parallel lines?

- What single transformation produces the same result as consecutive reflections over two intersecting lines?

Practice 4

1 Use squared paper to graph the image of each shape using the transformation given.

a Reflection in $y = -2$

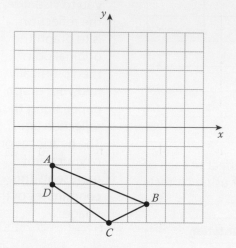

b Reflection in the x-axis

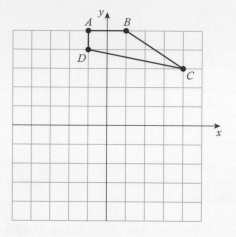

c Reflection in $y = -x$

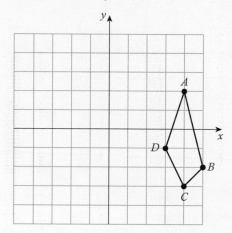

d Reflection in $y = -1$

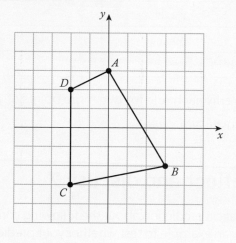

2 Write a rule to describe each transformation.

a

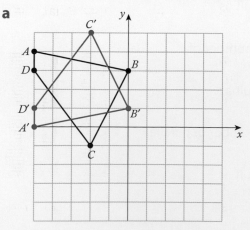

b

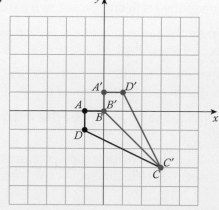

▶ Continued on next page

c

d

3 Give the coordinates of the image point after each of the following preimage points has been reflected in the *x*-axis.

    a (5, 9)   **b** (−2, 0)   **c** (0, −8)

4 Give the coordinates of the image point after each of the following preimage points has been reflected in the *y*-axis.

    a (−6, 15)   **b** (13, −11)   **c** (0, 0)

5 It seems that even superheroes love to use transformations to create their logos.

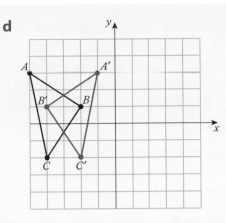

  a For each of the logos shown here, describe the transformations that have been used to create it. Copy each logo and draw in any reflection lines and indicate any center and angle of rotation where appropriate.

  b The above logos are based on transformations. What is it about transformations that produces logos that are pleasing to look at?

6 This is a design from a Navajo Native American rug. While there are many interpretations of the pattern, most Navajo grandmothers will tell you that the small diamond shape symbolizes the Navajo homeland, with each corner of the diamond representing one of the four sacred mountains that surrounded it.

Identify the original shape(s), describe the transformations that have occurred and explain how the shapes are tessellated.

▶ Continued on next page

7 This carved wooden screen window from Uzbekistan, with its geometric patterns, avoids the depiction of living things, as required by the Muslim faith. It also expresses the religious requirement to avoid using costly materials, so artistic pieces were often made of wood, clay or brass.

    a Identify the original shape(s), describe the reflections that have occurred and explain how the shapes are tessellated.

    b Describe how the pattern could have been created with translations.

    c Describe how portions of the pattern could have been created using rotations. Indicate the center of rotation as well as the angle and direction of rotation.

8 Write down a single transformation that achieves the same result as:

    a a reflection in the x-axis followed by a 180-degree rotation

    b a reflection in the line $y = -x$ followed by a 90-degree rotation clockwise

    c a reflection in the line $y = x$ followed by a 270-degree rotation clockwise.

Formative assessment – Your personal brand

ATL2

As you have seen in this unit, a logo or a coat of arms can be used to identify a person, organization or nation. Many logos and coats of arms are used to express the ideals, beliefs, values and aspirations of an individual or group.

criterion **C**

    a Find some company logos and/or coats of arms that are different than the ones you have studied already to exemplify each of the congruence transformations: translation, reflection, and rotation.

    b Copy each logo/coat of arms into a document and annotate the image to show the type(s) of transformation it shows.

    c Design your own logo or coat of arms that expresses something important to you: some part of your identity, beliefs, values or aspirations. It must incorporate at least two of the three types of congruence transformation. Draw this logo on a grid and indicate important features, such as reflection lines, centers of rotation, description of translation, etc.

    d Use correct notation to describe each transformation.

    e Below the image, you must include an explanation of the significance of your logo/coat of arms and give details of the transformation(s) involved.

    f Create a small poster, using an app like Canva, with the examples you studied and your own personal logo/coat of arms.

Reflect and discuss 8

Reflect and discuss 8

Your teacher will display your work around the classroom. Answer these questions after you have seen everyone's work.

- A logo/coat of arms is a very personal image. Which logos/coats of arms seemed most personal to you? Explain.

- Can you identify the person who created each logo/coat of arms, simply by looking at it? Explain with an example from your class.

Analyzing tessellations

The congruence transformations you just learned about are at the heart of any tessellation. You will now analyze tessellations to figure out how and why that is.

Activity 5 – Cultural tessellations

1 Download the app called "KaleidoPaint" to a mobile device.

2 Select the "Symmetry" menu and you will be offered a number of different template designs to create tessellations. Work in pairs, or groups of no more than four, with each person using their own device. See if you can identify the transformations that allow each template to create a tessellation. To help you, turn on the "Grids".

3 Select the first template (rectangles) and draw a basic shape/design. Then select each of the different templates in turn to see how your shape is altered. The background grid will help you to identify the transformations.

4 Once you are have finished exploring the app, create a tessellation of your own that is based on the theme of "culture". Conduct a gallery walk of all the devices and see how the theme has been developed by different people in your group. Come together as a class, with students taking turns to hold up their design and explain how it expresses "culture".

Activity 6 – Islamic art

You briefly looked at this tiling pattern from the Alcazar of Seville on page 270. You will now construct two of the shapes in it and investigate how they can be used to create this tessellation.

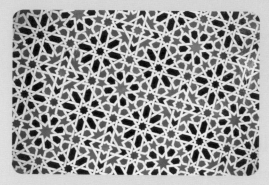

The larger star

1 On a piece of grid paper, draw x- and y-axes in the middle of the paper. Draw a giant circle centered at the origin. (Using a compass will make this easier.)

2 Using the same center as in step 1, draw a smaller circle inside your large circle.

3 Draw in a diameter from the top of the large circle to the bottom and another diameter from the left to the right of the large circle. Make sure the lines are perpendicular. Label these the x-axis and the y-axis.

4 Cut out the large circle, then fold the paper in half, and then again into quarters, along the axes.

5 Fold the paper further into an even number of sections or draw an even number of evenly spaced radii from the center of your circle to the circumference of the larger circle.

6 From the point where one fold/radius meets the circumference of the larger circle, draw a line to the point where the next fold/radius intersects the smaller circle. See the diagram below.

7 Then draw a line from this point to the point where the next fold/radius meets the larger circle.

8 Continue around the circle to produce a star.

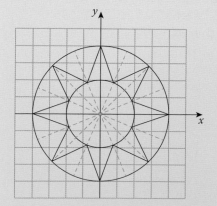

Keep in mind that this is an example. Your star may have more or fewer points.

▶ Continued on next page

9 Color in the star, making sure you can still see the folds or lines.

   a Describe any reflections in the x-axis.

   b Describe any reflections in the y-axis.

   c Could this shape have been created using a rotation? If so, define the center of rotation and the angle of rotation.

   d How could this star be modified so it could be tessellated to create a mosaic like the one in the photo at the start of the activity? Explain.

   e Paste your star onto another piece of paper and fill in around your star to create a base shape that can be tessellated. Explain why it is possible to tessellate this base shape and how you would do it.

The smaller star (8-pointed)

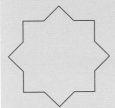

1 Create an 8-pointed star. The visuals below show each step of the process.

2 How could you describe these steps to somebody who did not have the visuals?

3 Color in the star, making sure you can still see the folds or lines. Indicate the x- and y-axes.

   a Describe any reflections in the x-axis.

   b Describe any reflections in the y-axis.

   c Could this shape have been created using a rotation? If so, define the center of rotation and the angle of rotation.

   d Can this star be tessellated to create a mosaic like the one at the start of the activity? Explain.

Reflect and discuss 9

- Do you find it easier to follow written instructions or visual instructions? Why? How could you use this information to help you to become a more efficient and effective learner?

- What makes this style of art so unique and distinct from other styles of art? Explain.

Investigation 3 – Shapes that tessellate

You will be looking at regular polygons and determining whether they will tessellate. You can do this using manipulatives (a set of regular polygons) or virtual manipulatives.

If you do not have a set of regular polygons to use for this investigation, you can use the Tessellation Creator on the NCTM website illuminations.nctm.org (see Activity 1 on page 260).

1 Determine which of the following regular polygons tessellate by themselves (without needing another shape to fill gaps). Copy and complete this table.

Regular polygon	Does this tessellate? (yes or no)
Equilateral triangle	
Square	
Regular pentagon	
Regular hexagon	
Regular heptagon	
Regular octagon	

2 Of the regular polygons that you tried, which ones tessellated?

3 What are the sizes of the interior angles of the shapes that tessellated?

4 How many degrees make up a full rotation?

5 Looking at your results in steps 4 and 5, what pattern can you find in the interior angles of the regular polygons that tessellate? What allows them to tessellate?

▶ Continued on next page

6 Can you predict which regular polygons will tessellate by themselves? Copy and complete this table.

Number of sides	Degree of interior angle	Number of polygons needed to fill a full rotation
3		
4		
5		
6		
7		
8		
9		
10		
11		
12		
n		

7 Given the formula you found in the last row of the table, test to see if a 20-sided figure would tessellate.

8 Verify your formula for two more polygons.

9 Justify why your formula works.

You have found the regular polygons that tessellate by themselves, but what if you combine two or more shapes together to complete the rotation? Tessellations of two or more different regular polygons such that the same polygon arrangement exists at every vertex are called *semi-regular tessellations*.

When you name these types of tessellations, you name them by the number of sides, starting with the polygon with the smallest number of sides and moving in a clockwise direction.

In a semi-regular tessellation, two or more shapes are used to complete a tessellation with no gaps, hence the vertices of the shapes connect and the shapes fit together to complete a full 360-degree rotation.

Activity 7 – Semi-regular tessellations

There are eight semi-regular tessellations.

1 Your task is to find all eight different semi-regular tessellations. For steps a–c:

- determine the combinations

- test them using manipulatives (real or virtual)

- sketch a full rotation of each and explain why it tessellates.

There are:

a three combinations of regular polygons that will tessellate with exactly three shapes put together (two of which may be the same shape)

b two combinations of regular polygons that will tessellate with four shapes (two of which may be the same shape)

c three combinations of regular polygons that will tessellate with five shapes (up to four of which may be the same shape).

2 Why do we only consider regular polygons and not irregular polygons?

You have looked at tessellations in Islamic art, such as those in the Alhambra. Tessellations appear in a wide range of art forms, such as the works of M.C. Escher (1898–1972), who is most famous for his "impossible" constructions and his tessellated works of art. Escher wanted to express both the mathematics of space and the logic of space in his artwork. Instead of being geometric shapes, his tessellating tiles look like recognizable things, such as a bird, a car, a dancer or a fish. However, he applied the same underlying principles as with regular polygons.

Activity 8 – The art of M.C. Escher

Below is a piece Escher created called "Bird No.44", which he completed in 1941 in ink and watercolor. If you look closely at the point where three beaks of the same color meet, you can see a rotation and can count six shapes completing the rotation. This makes the tessellating shape an equilateral triangle which is tessellated using a rotation.

Research Escher's tessellations and select **three** works of art to analyze. Identify the base polygon shape that was tessellated and describe how it was tessellated.

Reflect and discuss 10

- Do you like this style of art? Why or why not?
- Why do you think geometry was used so much in this style of art?
- Do you think that if you saw a piece by Escher you would recognize it? What would you use to identify it?

Similarity transformations

Not all transformations produce images that are the same size as the preimage. Sometimes, when a shape is transformed, it produces an image that is larger or smaller than the preimage. These images are *similar* rather than congruent. Transformations that produce similar figures are called *similarity transformations*.

Dilations

A *dilation* (also known as an *enlargement*) is a transformation that changes the size of the shape. The *scale factor* determines whether the dilation is an *enlargement* or a *reduction* and how many times larger or smaller the image will be compared with the original pre-image. A dilation also needs a *center point* to be defined: this is the point from which the dilation is made. Dilations are the only type of transformation that changes the size of the original shape; all the other types of transformation are simply **movements** of the preimage.

Investigation 4 – Dilations

ATL2 You will be using dynamic geometry software, such as the Polygon Transformer app on the ct.excelwa.org website, to complete this investigation.

1 Create a triangle A (1, 2), B (3, 3), C (5, 1).

2 From the transformations menu, construct a dilation image with center point (0, 0) and a scale factor of 3.

3 Copy and complete this table.

	Preimage triangle	Image triangle	Relationship between preimage and image
Vertex A	(1, 2)		
Vertex B	(3, 3)		
Vertex C	(5, 1)		
Horizontal distance A to B			
Horizontal distance A to C			
Horizontal distance B to C			
Vertical distance A to B			
Vertical distance A to C			
Vertical distance B to C			
Area of triangle			

4 Create a new table and repeat steps 1–3 using a scale factor of $\frac{1}{2}$.

5 What kind of scale factors produce enlargements and what kind produce reductions?

▶ Continued on next page

6 Generalize your results from steps 1–4 to explain how dilations are created using the origin as the center of dilation.

7 Verify your rule for two new points and a scale factor of your choosing.

8 Justify why your rule works.

9 Construct three lines that pass through the origin and each of the vertices of the original triangle. What do you notice? Explain why this makes sense.

10 How could you use these lines to determine the location of other dilations of the original triangle using the origin as the center?

11 How could you write dilations using arrow notation?

12 How would you write the examples you have studied in this investigation using arrow notation?

Reflect and discuss 11

- What if the origin was not the center point of the dilation? Move the center point to see what happens to the original triangle and the dilated image.

- Use the relationships you found in the investigation to write step-by-step instructions on how to create a dilation image regardless of the location of the center point and the scale factor of the dilation.

Example 3

Q A shape is to undergo a dilation of scale factor 5 with the center at the origin. If point A of the shape is at $(-2, 9)$, what will the coordinates of the point A' be?

A $(-2, 9) \rightarrow \times 5 \rightarrow (-10, 45)$

The coordinates of A' will be $(-10, 45)$.

> The dilation is centered at the origin, so you can find the image point by multiplying the ordered pair by the scale factor. In this case, the scale factor is 5.

Q What is the scale factor of the dilation (with center at the origin) if point A $(16, 36)$ becomes A' $(4, 9)$?

A $4 \div 16 = \dfrac{4}{16} = \dfrac{1}{4}$

so the scale factor is $\dfrac{1}{4}$.

> The dilation is centered at the origin, so you can find the scale factor by dividing the x-coordinate of the image by the x-coordinate of the preimage.

▶ Continued on next page

Check:

$$36 \times \frac{1}{4} = \frac{36}{4} = 9$$

Verify the scale factor by multiplying the original y-coordinate by the scale factor. If your answer is correct, you should get the y-coordinate of the image.

Q Draw the image of triangle ABC after a dilation with a scale factor of 2, centered at the origin.

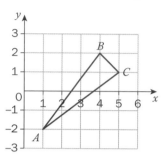

A $(1, -2) \rightarrow \times 2 \rightarrow (2, -4)$

$(4, 2) \rightarrow \times 2 \rightarrow (8, 4)$

$(5, 1) \rightarrow \times 2 \rightarrow (10, 2)$

The dilation is centered at the origin, so you can find the image by multiplying the ordered pair of each vertex by the scale factor. In this case the scale factor is 2.

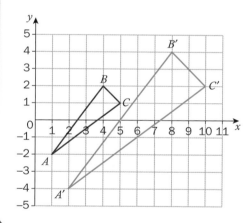

Plot the points of the vertices of the image triangle and label them correctly. Join them with straight lines.

Just like the other transformations, dilations that are on the Cartesian plane can be represented by specific notation. You will explore this notation in the next activity.

Activity 9 – Dilations on the Cartesian plane

1 Complete this example of correct notation in order to show a dilation with a scale factor of 2, centered at the origin.

$(x, y) \rightarrow (\underline{\quad}, \underline{\quad})$

2 Create two dilations of your own of a four-sided shape on squared paper. Include one reduction and one enlargement, both centered at the origin.

3 Describe each of your dilations using correct notation and find the location of each of the dilated images.

Practice 5

1 Using a dilation of scale factor 4 with the center at the origin, find the coordinates of the image A' of the point A (3, −7).

2 Using a dilation of scale factor $\frac{1}{3}$ with the center at the origin, find the coordinates of the image B' of the point B (−12, 27).

3 Using a dilation of scale factor $\frac{4}{5}$ with the center at the origin, find the coordinates of the image C' of the point C (20, 0).

4 Find the scale factor of the dilation (with center at the origin) if point D (9, 13) becomes D' (27, 39).

5 a Find the scale factor of the dilation (with center at the origin) if point E (48, −20) becomes E' (12, −5).

b What do you think a negative scale factor indicates? Explain.

6 Find the scale factor of the dilation (with center at the origin) if point F (0, −70) becomes F' (0, 21).

7 Rewrite questions 4–6 using correct notation.

► Continued on next page

8 Graph the image of the quadrilateral *ABCD* after a dilation with a scale factor of 3, centered at the origin.

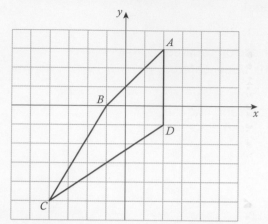

9 Graph the image of the triangle *ABC* after a dilation with a scale factor of $\frac{1}{3}$ centered at the origin.

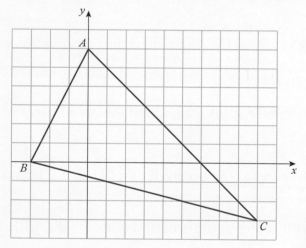

10 This is the logo for the Subaru car company. "Subaru" is the Japanese name for the cluster of stars known as the Pleiades in English (part of the constellation Taurus). "Subaru" also means "united" in Japanese – this expresses the idea that five companies merged to form Fuji Heavy Industries (FHI), the parent company of Subaru. So the star logo represents both a play on words and how FHI considers itself a united cluster of companies.

▶ Continued on next page

a Describe any congruence transformations that are visible in this logo.

b Describe any similarity transformations that are visible in this logo.

11 This pattern is based on traditional Aztec print designs.

a Identify the original shapes and describe all the transformations you can see in the design.

b Would you consider this a tessellation? Justify your response.

12 This is an example of a fractal design. In math, a fractal is an abstract object formed by a never-ending pattern repeated at increasingly small scales.

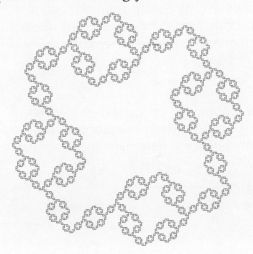

a Describe the dilations that you see in the fractal design.

b Could this shape be tessellated? Explain.

Go to nrich.maths.org and search for "Transformation Game". You can play in groups of two to four. You will need to download and print the game board and cards. You will also need to cut out a triangle for each player to use as a counter. Then follow the "How to play" instructions to play the game.

Activity 10 – Creating a transformation puzzle

For this activity you will need to be in groups of three.

Creating the visual representation

1 On a Cartesian plane with x- and y-axes no longer than −10 to 10, draw a quadrilateral and label the vertices.

2 Below the Cartesian plane, list written instructions for four different transformations: a translation, a reflection, a rotation and a dilation. Be sure to include all of the information that someone would need to carry out each transformation.

Creating the algebraic representation

3 On a separate sheet of paper, list the four ordered pairs of the vertices of the quadrilateral.

4 Represent each transformation using arrow notation.

Round 1

5 Pass your visual representation and written instructions to the person on your right. They will carry out the transformations in order, applying each transformation to the result of the one before and drawing each resulting image. They will then color the final image.

Round 2

6 Pass your algebraic representation to the person on your left. They will draw the original shape and the image after each transformation, using only algebra to determine the coordinates of each image.

7 Compare your answers and check for accuracy.

ATL1

Reflect and discuss 12

Discuss these questions with your group.

- Compare using the visual representation and the algebraic representation. Which process did you find easier? Explain.

- Do you need to know how to do both methods or could you rely on just one? Explain.

- How does knowing how to use multiple methods to solve a problem make you a more effective learner? Explain.

297

Unit summary

A *tessellation* is the tiling of a surface using geometric shapes with **no** overlaps and **no** gaps.

A *transformation* changes an original shape, called the *preimage*, and produces a shape called the *image*.

Translation, rotation and reflection are all *congruence transformations*. Congruence transformations create images that are identical in size and shape to the preimage.

Dilation is a *similarity transformation*. Similarity transformations create images that are identical in shape to the preimage and are proportional in size.

Translation – moves a point or shape in any direction with no change to size. If a shape is translated, every vertex of the shape must be moved in the same direction and by the same distance.

Translations can be represented using the notation

$$(x, y) \rightarrow (x + h, y + k),$$

where h and k are real numbers.

Reflection – flips a point or shape over a given line called the *line of reflection* or the *mirror line*.

Reflections can be represented using this notation:

$(x, y) \rightarrow (x, -y)$	a reflection in the x-axis
$(x, y) \rightarrow (-x, y)$	a reflection in the y-axis
$(x, y) \rightarrow (y, x)$	a reflection in the line $y = x$
$(x, y) \rightarrow (-y, -x)$	a reflection in the line $y = -x$

Rotation – turns a shape about a fixed point through a determined number of degrees. The center is called the *point of rotation*.

Rotations about the origin can be represented using this notation:

$(x, y) \rightarrow (-y, x)$	a 90-degree rotation clockwise
$(x, y) \rightarrow (-x, -y)$	a 180-degree rotation clockwise
$(x, y) \rightarrow (y, -x)$	a 270-degree rotation clockwise
$(x, y) \rightarrow (x, y)$	a 360-degree rotation clockwise

Dilation – creates a bigger or smaller image than the original (pre-image). The *scale factor* indicates the size change and the center of the dilation must be given. In the examples in this book, the center of the dilation is always the origin.

Dilations with the origin as the center of dilation and scale factor *a* can be represented using this notation:

$$(x, y) \rightarrow (ax, ay),$$

where *a* is any real number.

Scale factors larger than 1 produce enlargements of the preimage. Scale factors between 0 and 1 produce images that are smaller than the original (reductions). Negative scale factors produce images that are on the other side of the centre of dilation and are upside down.

Unit review

criterion **A**

> 📖 **Launch additional digital resources for this chapter**

Key to Unit review question levels:

Level 1–2   **Level 3–4**   **Level 5–6**   **Level 7–8**

1 Copy and complete the table below.

Point	Translation up 3 units	Translation left 5 units	Translation down 2 units and right 7 units	Dilation by a scale factor of 4 about the origin
(0, 0)				
(6, 1)				
(3, −2)				
(−4, 7)				
(−8, −5)				

2 **Plot** the image of each shape on your own copy of the Cartesian grid.

a Translation 4 units right and 4 units down

b Translation 2 units right and 3 units up

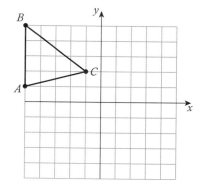

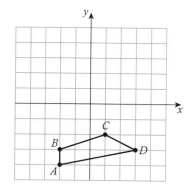

3 The photo shows a detail of the Serpent Bench in the Parc Güell, in Barcelona, in the Catalonia region of Spain. The bench winds around the park's perimeter and is covered in a mosaic of ceramic tiles. It was created by the artist Antoni Gaudí, who expressed his love of natural forms in his fantastical designs. Eusebi Güell, who commissioned the work, wanted Gaudí to represent the connection between ancient Greek symbols, Christian symbols and symbols of Catalonia.

Is this design a tessellation? **Explain** why or why not.

4 **Describe** the transformations that have occurred in each diagram.

a

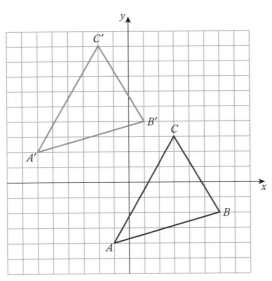

b

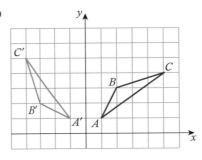

c

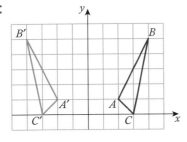

d

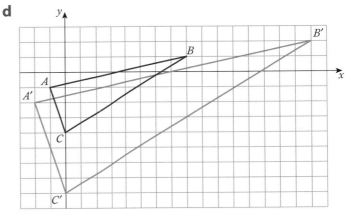

e

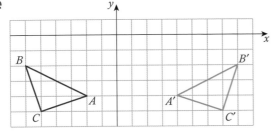

5 Using a dilation of scale factor 5, with the center at the origin, find the coordinates of the image L' of the point L (12, −9)?

6 What is the scale factor of the dilation (with center at the origin) if the image of point T (6, −18) is T' (2, −6)?

7 **Plot** the image of the shape with vertices (−5, 8), (3, −6) and (−3, 0) using the following translations.

a $(x, y) \rightarrow (x + 4, y - 9)$

b $(x, y) \rightarrow (x - 3, y + 6)$

8 Copy and complete the table below.

Point	Reflection in x-axis	Reflection in y-axis	Reflection in $y = x$	Reflection in $y = -x$	Reflection in $y = 3$	Reflection in $x = -4$
(0, 0)						
(8, 2)						
(7, −6)						
(−4, 1)						
(−5, −10)						
(0, −3)						
(9, 0)						

9 **Plot** the image of each shape on your own copy of the Cartesian grid.

a Rotation 90° clockwise about the origin

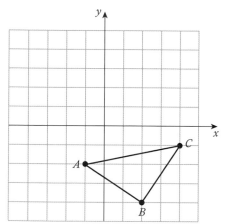

b Rotation 180° about the origin

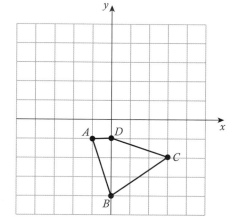

10 This is a folk pattern used in traditional Romanian embroidery. It is said that every form on a Romanian woman's blouse represents an aspect of the woman's history and family. Even the color of the pattern expresses the region where it was made and can indicate the age and social status of the woman wearing the blouse.

   a **Identify** the original shape(s) used to make this pattern.

   b **Describe** the transformations that have occurred to the original shape(s).

   c **Explain** how the shapes are tessellated.

11 Copy and complete this table. All rotations are about the point (0, 0).

Point	Rotation of 90 degrees clockwise	Rotation of 180 degrees clockwise	Rotation of 270 degrees clockwise	Rotation of 360 degrees clockwise	Rotation of −90 degrees clockwise
(0, 0)					
(8, 2)					
(7, −6)					
(−4, 1)					
(−5, −10)					
(0, −3)					
(9, 0)					

12 **Plot** the image of each shape on your own copy of the Cartesian grid.

   a Reflection in $x = -3$

                                    b Reflection in $y = x$

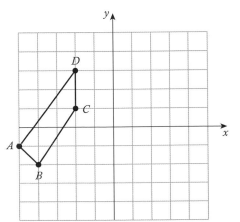

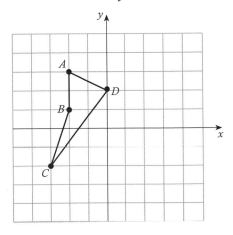

13 Describe the transformation that has been performed in each diagram and express it using correct arrow notation.

a

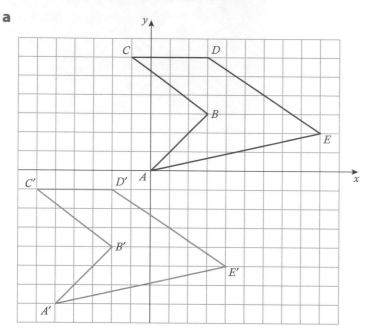

b

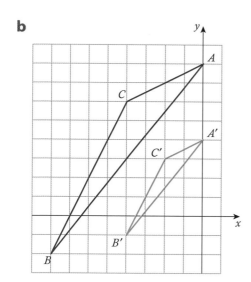

c

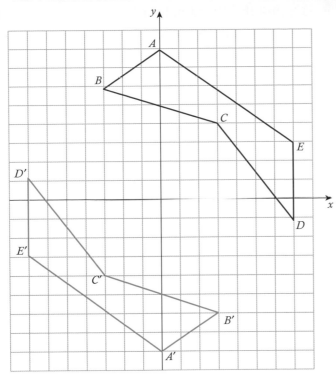

d

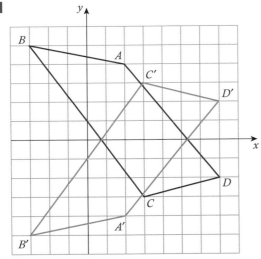

14 Given the rotation $(x, y) \to (y, -x)$, find the coordinates of the image of the given points and describe the rotation in words.

   a $(7, 14)$

   b $(-12, -5)$

   c $(0, 3)$

15 Given the reflection $(x, y) \rightarrow (-y, -x)$, find the coordinates of the image of the given points and **describe** the reflection in words.

a $(0, -9)$

b $(14, -4)$

c $(3, 5)$

16 This door is in the Forbidden City, a palace in Beijing, China. Philosophy and religion are expressed in the art and architecture throughout the Forbidden City, as is the power of the Chinese imperial family. Even colors express ideas and beliefs, with the color red symbolizing power, happiness, luck and wealth.

a **Identify** the original shape(s) in this design.

b **Describe** the transformations that have occurred to the original shape(s).

c **Explain** how the shapes are tessellated.

17 Is a reflection in the line $y = -x$ the same as a rotation of 180°? **Explain** using arrow notation.

18 In ancient Rome, "tesserae" were small squarish tiles that artists used to make bigger pictures that covered floors and walls. This mosaic is a geometric mosaic found in the ancient city of Ephesus, Turkey. Mosaic floors such as this one were used to express wealth and power, since most people could not afford such a luxury. The symbols within mosaics had their own meaning. Some mosaics are meant to depict important events or people. A mosaic like this one is full of symbolism, such as the Solomon's knot figure (the two intertwined bands) which often expressed a belief in immortality or eternity.

a Identify preimage shapes in the mosaic that have been transformed.

b Describe the transformations that have occurred to these preimage shapes.

c Explain how the shapes are tessellated.

19 What single transformation gives the same result as each of the following?

a A reflection in the *x*-axis followed by a reflection in the *y*-axis

b A reflection in the line $y = x$ followed by a 90-degree clockwise rotation about the origin

20 This Chinese pattern was used on cards, paper and fabric to depict the Chinese New Year of the Rooster in 2017.

a If the larger circles have a diameter of 5 cm, **identify** the preimage shape(s) and **describe** the transformations that were used to create the pattern.

b Describe another transformation that could have been used to create this design.

Did you know...?

There are 12 Chinese zodiac symbols, each represented by a different animal. A person's zodiac animal symbol is determined by the year in which were born. According to legend, the order of the animals was decided in the Great Race initiated by the Jade Emperor. The 12 animals, and therefore the people born under each sign, have their own symbolism and traits. The Chinese zodiac reinforces the belief that everyone (and every animal) has a role to play in life and society.

21 **Describe** the transformation that has occurred to create the image in the diagram below.

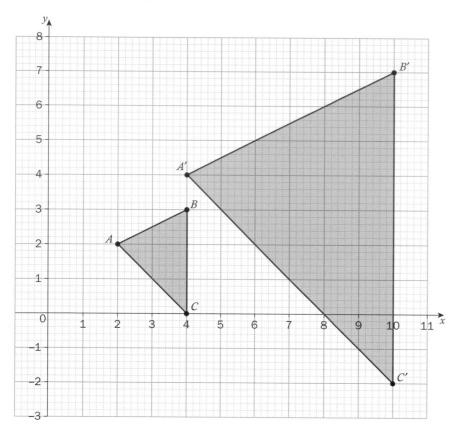

22 This is a wood panel at the Sejong Cultural Center in Seoul, Korea, made as a tribute to ancient panels that were used in traditional Korean houses. Many homes had such geometric designs as window frames and walls. Korean paper, made of mulberry bark, is very strong and was used instead of glass prior to the 19th century. The paper would be replaced every year or so.

a Select two panels that contain a tessellation that could be created using two different types of transformation.

b **Sketch** the tessellation in each of your chosen panels. **Annotate** your sketch to show the original shape and the geometric transformation that has occurred to create each tessellation.

23 This window is built of red sandstone and is part of an ancient monument called Humayun's Tomb in New Delhi, northern India. Built to honor the Emperor Humayun, the window uses geometric shapes much like those used in other Muslim buildings. The six-pointed star (or hexagram) is often associated with Judaism, so it may seem a surprising symbol to see here, but it is likely that it was used to honor and express Humayun's interest in astrology.

Describe a base shape in this tessellation. Then **explain** how this shape has been transformed and what other shapes have been added to create the tessellation

24 The original vertices of a triangle are (1, 0), (3, 4) and (6, 2). The vertices of the image of the triangle are (−6, −5), (−10, 3) and (−16, −1). Determine the three transformations that have occurred.

25 The original vertices of a triangle are (3, 2), (1, 9) and (−6, 4). The vertices of the image of the triangle are (9, −2), (7, −9) and (12, −4).

a Determine **two different sets** of two transformations that could have occurred.

b **Explain** how it is possible for two sets of different transformations to create the same overall combined transformation.

26 The original triangle has undergone three transformations (a rotation followed by a translation, followed by a reflection) to become the image triangle.

Identify the three transformations and **use** appropriate notation to describe them.

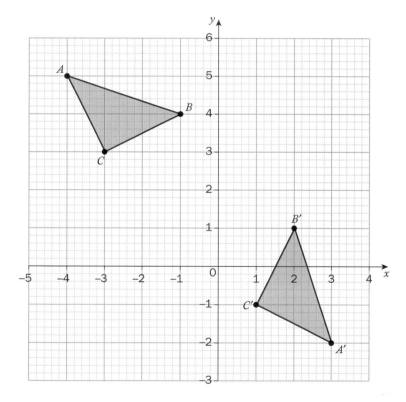

Summative assessment

Creating your own tessellation

As you have already discovered, a tessellation (or tiling) is when you cover a 2D (or even 3D) surface with a pattern of polygons so that there are no overlaps or gaps. You have done a lot of work analyzing patterns and learning the different ways in which shapes can be transformed (translations, reflections, dilations, rotations). Now it is time to create a tessellation for yourself that reflects you as a person.

> You will be using the tessellations.org website to guide you through this assignment. Select the 'Make Your Own' menu to see the different ways to create a tessellation. The website demonstrates a range of different methods for creating tessellations. You might like to try a few of the different methods before you choose the one you will use for your own tessellation. If you are short of time, each student in your group could study a different method and explain it to the rest of the group.

a Read the "All about tessellations" section and do the "Self-test: tessellation or not?" to give you some additional groundwork on tessellations.

b Go to the "Make your own" section and start by working through the "How to Make a Tessellation" topics. Focus on analyzing the following two methods:
  - line method
  - slice method.

c Your task is to create a tessellation using either of these methods (or another method described on the website) using dynamic geometry software (or by hand). The piece of art must be a representation of you in some way.

d Produce a report that includes:
  - a tessellation using a method explained on the website
  - a complete storyboard showing each step of the process in developing your tessellating shape (see details below), together with a mathematical explanation of each step
  - an explanation of how this tessellation is a representation of you.

Storyboard details

Section 1

- The original geometric shape – why this shape?
- A mathematical explanation of why this shape will work in this tessellation

Section 2

- The tessellated shape
- An explanation of the changes you have made to the geometric shape to create your tessellated shape (your tile)
- A description of all of the transformations that are visible in the tessellated shape:
  - rotation
  - translation
  - reflection
- A mathematical explanation of the exact transformations and why they work

> Be sure to indicate lines of reflection, centers of rotation and angles of rotation.

Section 3

- Your tessellation design – present this in color to fit on one A4 sheet. Your teacher will ask you to either print your design or upload it to a shared location. All of the printed student designs can then be attached together to form a "class quilt" that will hang in the school.

An example of the images for an e-storyboard

Shape	Mathematical explanation
1	

Shape	Mathematical explanation
2	
3	

Reflect and discuss

- Describe the benefits and limitations of using technology to create a tessellation.

- Geometric transformations and tessellations are found throughout history and span many cultures. Why do you think that is?

- Can an understanding of mathematics help improve creativity? Justify your response.

- How does your tessellation express your culture, beliefs and values? How is it a representation of you?

7 Linear systems

Making socially responsible choices is in itself a choice, but is it worth the associated extra cost? In this unit, you will discover how you can use a system of linear equations as a decision-making tool. These systems will help you analyse your options and see what you may be gaining or losing as a result. However, linear systems can also be used with a range of other decisions, from the truly personal to those that cause major events in history.

Identities and relationships

Personal health and values

How do you make decisions? What factors do you consider? Do you ever use mathematics to help you analyse your choices? Systems of linear equations can be very useful tools in selecting options that also reflect your personal values.

A healthy diet includes a combination of foods from a variety of food groups. Balancing all of the nutritional elements can be difficult, but the options could be represented by systems of equations to be solved. The solutions of these systems could guide your choices for the optimal diet based on your age and activity level.

Trying to decide how much time to spend on homework versus leisure activities? Wondering whether buying a car is better than leasing the same vehicle? Not sure if taking out student loans is a better option than working while in school? All of these situations lend themselves to systems of linear equations. Maybe mathematics should be the first place you turn to when making personal decisions!

Turning points in human history

"What would have happened if...?"

We sometimes ask this question as we reflect on major events that lead to turning points in our history. If the results of major decisions had been analysed before they were made, how might history have turned out differently?

In 1920, Prohibition became law in the United States and the production, sale and transportation of alcohol became illegal. The goal was to reduce crime and the amount of money spent managing prisons and prisoners. Individual states expected to make huge financial savings. However, the new law, which was in effect for 13 years, created as many problems as it solved. Prohibition encouraged a new kinds of criminal activity (making and distributing contraband alcohol), reduced overall income and led to corruption in some areas of law enforcement.

Even Napoleon Bonaparte, considered to be one of the greatest generals in history, could have used mathematics to his advantage. He is seen to have made a number of bad decisions, from how he invaded Russia, to the battle at Waterloo. When given several options, linear systems could have helped him see when one choice would have been better than another.

Human history is full of decisions, some great and others not so great. Systems of linear equations can be used to analyse the choices that were available and help us to decide whether or not the right decision was made. How might the world have been different with a little more mathematics?!

7 Linear systems
Social entrepreneurship

KEY CONCEPT: RELATIONSHIPS

Related concepts: Representation, Models

Global context:

In this unit, you will use linear systems to model relationships as you explore a relatively new idea, *social entrepreneurship*. As part of the global context of **fairness and development**, you will see how people are trying to make a difference in the lives of others through the companies they start and the products they sell. Along the way, these entrepreneurs need to use tools, such as linear systems, to help them make the important decisions that affect their ability to sustain their mission.

Statement of Inquiry:

Representing relationships with models can promote and support social entrepreneurship.

Objectives

- Solving complex multi-step algebraic equations
- Representing and classifying systems of linear equations
- Solving a system of linear equations using graphing, substitution and elimination
- Applying mathematical strategies to solve problems using a system of linear equations to help in decision-making

Inquiry questions

F
What is a relationship?
What does it mean to "break even"?

C
How are relationships represented with models?

D
What is our responsibility to those in our community and other communities?
How can I make a difference?

You should already know how to:

1 Find the degree of a polynomial:

State the degree of each polynomial.
$2x^2 + 3x - 5$ $4x^3 - 5x^4 + x^2 - 8$
$2x^2y + 6xy^2 - 7x^3y^2$

2 Solve equations

Solve the following equations.
$3(x - 2) + 5 = -4$ $\dfrac{x}{2} - 3 = 1$ $x^2 = 16$

3 Draw linear functions

Sketch the following lines.
$y = -2x + 7$ $2x - 3y = 6$ $4x + y - 2 = 0$

4 Write the equation of a line in standard form and gradient–intercept form

Write each equation in standard form and gradient–intercept form.
$4x - y = -12$ $3x - 4 = -2y$ $x = -5y + 9$

5 Find x- and y-intercepts

Find the x- and y-intercepts for each function in question 4.

6 Find the slope of a line using the formula

Determine the slope of the line through each pair of points.
$(-1, 3)$ and $(2, 5)$ $(-2, -2)$ and $(3, -1)$
$(0, -7)$ and $(2, -1)$

7 Find the slope of a line parallel and perpendicular to a given line

Find the slope of a line that is parallel to each of the lines in question 4.

Find the slope of a line that is perpendicular to each of the lines in question 4.

8 Find the equation of a line

Find the equation of the line with the following characteristics.

a gradient of -2 and y-intercept of 4

b gradient of $\dfrac{1}{4}$ and a y-intercept of -8

c passes through the points $(-1, 2)$ and $(3, -2)$

Introducing linear systems

What goes into the decisions that people make? Is cost always the most important factor or are there other elements that need to be considered? For instance, would you be willing to pay more for a product that was created using environmentally friendly materials? How important is it that the company you buy from uses a workforce that is well paid and has good working conditions? Would you prefer to buy from a company that is using its profits to solve a social issue? Do you only support local companies?

In this unit you will analyze decisions by creating simple models that consider some of the important factors involved. These same types of models are used every day by individuals, companies and governments when they make decisions.

By the end of this unit, you may be ready to start your own company and dedicate your profits to a worthy cause of your choice!

Reflect and discuss 1

- When you shop for products, what factors affect your final choice?

- Give an example of a time when you had to choose among several options where your decision would directly affect one or more people. How did that influence your decision-making process?

Solving linear equations

Solving equations is a very important skill in mathematics. However, there are different kinds of equations, which may require different strategies to solve.

Investigation 1 – Types of equations

Pairs

Perform the following investigation with a peer.

1 Copy and complete this table.

Equation	Degree	Classification
$3g + 2 = 12$		linear
$4m^2 - 8m + 3 = 0$		quadratic
$c^3 - 4c = 2$		cubic
$\dfrac{4x - 5}{7} = 9$		
$12 - 4w^3 + 2w^2 - w = 2$		
$3y + 11 = -7y^2$		
$6z - 19 = 2(3z - 4)$		
$5p^3 - 2p + 4 = -2p^2 + p^3 - 8p$		
$7h^2 - 9h = -12(3h + 2)$		

2 Based on your results, define the following terms.

- linear equation
- quadratic equation
- cubic equation

3 What do you think a "quartic equation" is? Give an example.

4 Does the choice of variable (x or m, etc.) affect the classification of the equation? Explain.

In this unit, you will be solving linear equations. You may have noticed that each equation in the investigation contained only one variable. When there is only one variable, a single equation is sufficient to find the value of that variable (or "solve the equation"). You have already learned strategies to solve different types of equations. You will apply those same strategies here to solve more complex equations.

Example 1

Q Solve the following equations.

a $\dfrac{x}{6} = 15 - \dfrac{2x}{3}$ **b** $x^2 + 75 = 100$ **c** $\dfrac{2(x+7)}{5} = 2x + 6$

A **a** $\dfrac{x}{6} = 15 - \dfrac{2x}{3}$

> If you have a fractional equation, find a common denominator and multiply every term in the equation by that common denominator.

$\dfrac{\cancel{6}x}{\cancel{6}} = 6(15) - \dfrac{2\cancel{6}(2x)}{\cancel{3}}$

> Simplify.

$x = 90 - 4x$

$x + 4x = 90 - 4x + 4x$

> Rearrange the equation by doing the opposite operation to isolate the variable.

$\dfrac{5x}{5} = \dfrac{90}{5}$

> You have finished when you have isolated the variable on one side.

$x = 18$

b $x^2 + 75 = 100$

$x^2 + 75 - 75 = 100 - 75$

> Simplify and rearrange the equation by doing the opposite operations to isolate the variable.

$x^2 = 25$

$\sqrt{x^2} = \sqrt{25}$

> You have finished when you have isolated the variable on one side. Note that, when finding the square root of a number, you must take both the positive and negative answers into account.

$x = \pm 5$

c $\dfrac{2(x+7)}{5} = 2x + 6$

> When you have brackets in the equation, apply the distributive law, so that every term on the inside of the brackets is multiplied by every term outside of the brackets.

$\dfrac{\cancel{5} \times 2(x+7)}{\cancel{5}} = 5(2x + 6)$

$2x - 2x + 14 = 10x - 2x + 30$

> Simplify and rearrange the equation by doing the opposite operations to isolate the variable.

$14 - 30 = 8x + 30 - 30$

$\dfrac{-16}{8} = \dfrac{8x}{8}$

$x = -2$

> You have finished when you have isolated the variable on one side.

Reflect and discuss 2

- Classify each equation in Example 1 (linear or quadratic). Justify your answers.

- How is the number of solutions related to the degree of the equation? Explain.

- With a peer, review each of the examples and help each other to understand the strategies that were used.

Practice 1

1 Solve each equation.

a $3(x+4) = -3x$

b $\dfrac{5x}{2} = 4x + 3$

c $x - 4 = \dfrac{x}{4} + 20$

d $\dfrac{4(x-5)}{5} - 3x = x - 36$

e $\dfrac{2x-3}{6} - \dfrac{x}{4} = 1$

f $\dfrac{x-5}{4} = 13 + \dfrac{3-4x}{9}$

g $\dfrac{6x+6}{3} = \dfrac{4x+6}{4}$

h $2x^2 + 100 = 5x^2 - 407$

i $\dfrac{4x-9}{3} - \dfrac{5x+12}{6} = 3x$

j $\dfrac{2}{3}x - \dfrac{4}{3} = \dfrac{1}{2}x - \dfrac{3}{2}$

k $7x^3 - 12x + 5 = 4x^3 + 9x - 11 + 3x^3$

l $4x^2 - 11 = 2x^2 + 9 - 3x^2$

2 Create a linear equation that requires multiple steps to solve where the answer is −4.

3 Create a fractional linear equation that requires multiple steps to solve where the answer is 2.

4 Create a quadratic equation that requires multiple steps to solve where the answers are 7 and −7.

Did you know...?

Most companies are "for profit" enterprises, which typically measure performance using profit and revenue data. The ultimate goal is to increase the value of the company. *Social entrepreneurs* run a business with the primary goal of finding a solution to a social or environmental issue. With the help of the internet and social media, they can reach many people and ask for support to help implement their ideas for change.

For example, believing that credit is a fundamental right, Muhammed Yunus created the Grameen Bank. His bank provides small loans (called *microloans*) to impoverished individuals and families in his native Bangladesh so that they can start a small business. Nearly 9 million people have borrowed money from the bank; over 97% of them have been women.

Solving systems of linear equations

You know how to solve equations with one variable, but what if there is more than one unknown? What if there are two variables? A *system of equations* is a group of equations that can be solved for each of the variables in it. With two variables, you will need at least two equations in order to find the values of both variables. If these two equations are both linear, you have what is called a *system of linear equations* or simply a *linear system*. You can solve this linear system using a variety of methods; these methods are the focus of this unit.

Solving linear systems by graphing

One of the methods that can be used to solve a linear system is graphing. Here, you will apply the skills you have learned in graphing lines on the coordinate plane, but now there will be two lines instead of just one.

Activity 1 – Graphing linear systems

1 On the same coordinate grid, draw the following lines.

$$y = 2x - 4 \text{ and } y = -3x + 1$$

2 Why do these lines intersect? Justify your answer mathematically.

3 Find the coordinates of the point where the two lines cross (the *point of intersection*).

4 Verify that the point of intersection is on both lines by substituting the coordinates of this point into the original equations.

5 Explain why the point of intersection solves the linear system.

6 Repeat steps 1 through 5 using the following equations.

a $y = -5x + 8$ and $2x + y = 2$

b $y = 3 - 2x$ and $4y - 7 = -6x$

Reflect and discuss 3

Pairs

Answer the following questions in pairs.

- What are the different ways of graphing a line such as $3x - 6y = 12$? Which method do you prefer?

- Describe any disadvantages you see to solving a system of linear equations by graphing.

- Why is it a good idea to verify your answer when solving a linear system? Explain.

- Is it possible for a linear system to have no solution? Explain.

 If you do not have access to a graphic display calculator (GDC), you can use an online graphing tool such as Desmos to graph each of the lines. Visit www.desmos.com to use this graphing utility. Be sure to write your equations in gradient–intercept form.

Activity 2 – Profit or loss?

An artist in Colombia makes macramé cotton hammocks like the one pictured here. The cost for the equipment needed to make the hammocks is $500 and the cost of materials for each hammock is $25. He decides to take out a microloan from a community development bank similar to Grameen Bank in order to start a business which he hopes will support his family. The selling price of a hammock is $70.

Let y be the amount in dollars and x be the number of hammocks sold.

1 *Revenue* is the amount of money a company takes in from sales of a product. Explain why the revenue from the hammocks can be given by the equation $y = 70x$.

2 Explain why the cost of making the hammocks can be given by the equation $y = 25x + 500$.

3 Solve the linear system by graphing both lines on the same coordinate grid.

4 The point of intersection of a system that represents a company's costs and revenue is called the *break-even point*. Why do you think this is?

5 To make a profit, the artist's revenue needs to be more than his costs. How many hammocks does he have to produce before he begins to make a profit? Explain your answer.

6 The same hammock on the North American market would cost the customer $120? Why do you think there is a difference in price? Is this unfair and, if so, to whom?

Practice 2

1 Rewrite each equation in gradient–intercept form.

a $2y = -4x + 8$

b $5x - 10y = 20$

c $0 = 2x - 3y + 9$

d $\frac{1}{2}y - 3x = 5$

e $\frac{2}{3}y = 2x + 6$

f $3y = \frac{1}{2}x - 12$

g $5x = 10y - 20$

h $x = 8y$

i $\frac{1}{2}x = 3y - 2$

j $\frac{2}{3}y - 8x = -6$

2 Find the x- and y-intercepts of the following linear equations.

a $3y + 4x = 20$

b $5x - 10y - 30 = 0$

c $0 = 2x - 4y + 12$

d $\frac{y - 3x}{2} = 6$

e $\frac{1}{4}y = 2x + 4$

f $4y = \frac{1}{2}x - 12$

3 Solve each linear system by graphing. Verify your answers.

a $y = 2x - 6$ and $y = x + 3$

b $y = -x + 2$ and $y = 5x - 4$

c $y = -\frac{1}{2}x - 4$ and $y = \frac{1}{3}x + 1$

d $y = -2x - 10$ and $y = \frac{3}{2}x - 3$

e $y - 7 = 6x$ and $3y + 12x - 6 = 0$

f $3x + 2y = 16$ and $2x + 3y = 14$

g $5y - 35 = -5x$ and $0 = \frac{3}{2}x - 3 - y$

> You can verify your answers by substituting the coordinates back into the equations.

4 A video streaming company is putting together their pricing strategy and is considering a couple of new promotions. Two of them involve a loyalty program where some of the proceeds will go to providing internet access to local students who do not have the financial resources to afford the internet. The current average rental cost is $6 per movie. The loyalty program options are: you can either download as many movies as you want for a $35 monthly fee or you can pay $10 a month to join the loyalty program and then the rental cost is just $4 per movie.

▶ Continued on next page

a Create a table of values for each of the three options that shows the cost of renting from 1 to 5 movies.

b Graph these options on the same set of axes.

c The results of a marketing survey showed that, on average, customers download 5 movie rentals each month. Which promotion is the better deal for this number of rentals? Explain.

d Write a general statement about when each option is more economically attractive to the customer depending on the number of movie rentals downloaded.

e How could such promotions increase a company's ability to raise money for their cause? Explain.

Number of solutions to a linear system

The systems of linear equations that you have solved so far all produced one solution. Is it possible for there to be more than one answer? Could there ever be no solution? Graphing systems is an efficient way of investigating the answer to these questions.

 Investigation 2 – Classifying systems of equations I

Here are three linear systems.

A	B	C
$y = 2x + 1$	$2y + 8 = x$	$6x = 20 - 2y$
$y = -x + 7$	$4y - 2x = -16$	$y + 3x - 5 = 0$

1 Graph each pair of lines on a separate set of axes. With the permission of your teacher, you may use technology if you wish.

2 How many solutions does each linear system have? Explain your answers.

3 Research the names used to classify the three types of linear systems.

▶ Continued on next page

4 Is it possible to determine how the lines will intersect before actually trying to find the solution by graphing? Rewrite all the above equations in the form $y = mx + c$ and copy and complete the table below.

	A	B	C
Linear system in $y = mx + c$ form			
Slope of each line			
y-intercept of each line			
Number of solutions			

5 Write a brief summary of the results of your investigation into the three types of linear systems.

6 Verify your results for at least two more linear systems.

7 Justify why your results make sense.

Practice 3

ATL2 Answer question 1 in pairs. If someone needs help, ask leading questions instead of giving answers.

Pairs

1 Without graphing, classify each linear system based on the number of solutions. Justify your answers mathematically.

a $y = 3x - 5$, $2x - 6y = 8$

b $y = 2 - \dfrac{3}{4}x$, $3x + 4y = -2$

c $4x = 6y + 1$, $9y = 6x - 5$

d $12 - 2x = 4y$, $5x + 10y - 30 = 0$

e $y = \dfrac{1}{3}x - \dfrac{5}{6}$, $6y - 2x + 5 = 0$

f $x = -1$, $y = 3$

g $4x - 2y = 14$, $12x - 6y = 40$

h $-x + 2y = 5 - y$, $-x + 3y = 5$

i $y - 3 = 4x$, $6 - 2y = 8x$

j $y = 2x - 2$, $6x - 3y = 2$

k $y + 3x = 5$, $5x + 1 = 3y$

l $x + y + 4 = 0$, $4x + 4y + 8 = 0$

▶ Continued on next page

2 Given the linear system

$y = 2kx + 6$

$4x + y = 2,$

a Find the value of k that makes the linear system have no solution.

b What values of k will result in one solution? Explain.

3

The YMCA is a non-profit, worldwide organization operating facilities in 119 countries. YMCAs differ from country to country and offer vastly different programs in response to local community needs. Local YMCAs raise money so that they can engage in a wide variety of charitable activities, including providing classes and mentorship for young people who want to become social entrepreneurs.

A local YMCA has a membership offer that charges youths a $90 one-time joining fee and then $0.75 for every day that the facilities are used. Alternatively, a day pass (which costs $5 each day) can be bought to use the facilities.

a Set up a table of values to represent both options.

b Graph the system.

c If you think you will use the facilities 6 times a month, how long will it take to pay off the membership fee and for this to be the better option?

d If you ended up only using the facilities once a week (4 times a month), would signing up for a membership be a good idea? Explain.

e Can you think of another reason why you would only use the day pass option and not sign up for a membership?

▶ Continued on next page

   f Sometimes YMCAs run promotions and offer a significantly lower joining fee. What if the joining fee dropped to $1 and you plan on going once a week? Would you sign up for the membership then? Justify your answer by solving the system of equations.

   g How would the problem have to be modified in order for there to be no solution?

4 You are given the system of equations $3x + 2y = 2$ and $-6x + 4y = -4$.

   a Classify this system. Justify your answer.

   b Verify your classification by graphing the system.

   c Change one term in the equation $-6x + 4y = -4$ so that the system has an infinite number of solutions. Explain why this works.

   d Show that your answer is correct by graphing the system with your new equation.

5 Create a linear system where there is no solution but this is not immediately obvious.

6 Create a linear system where there are infinitely many solutions but this is not immediately obvious.

While graphing gives you a good visual representation of the linear system, answers that are decimals or fractions may be difficult to determine. Sometimes it is easier and more efficient to solve the linear system algebraically. There are two methods commonly used – substitution and elimination.

Solving linear systems by substitution

ATL1

Pairs

Activity 3 – The substitution method

In pairs, complete the following activity.

1 Given the linear system

$$3x - 2y = 8 \qquad \text{①}$$
$$y = -2x + 3 \qquad \text{②}$$

   a substitute the expression for y from equation ② into equation ①

   b solve this linear equation for x

▶ Continued on next page

    c substitute the value for x into equation ② to find the value of y

    d write your answer as an ordered pair (x, y).

2 Given the linear system

    $4x + y = -2, -2x - 2y = 4,$

    a Which variable would be easiest to isolate in one of these equations? Explain.

    b Isolate this variable in one of the two equations.

    c Substitute the expression for this variable into the other equation, as seen previously.

    d Solve the linear equation.

    e Find the value of the other variable.

3 Summarize how to solve a linear system using the substitution method.

Reflect and discuss 4

Pairs

In pairs, answer the following questions.

- In question 1, what made equation ② easy to substitute into equation ①?

- What would you have to do if equation ② were written in standard form $(Ax + By + C = 0)$?

ATL2 Can you identify any disadvantages of the substitution method? Discuss these with a peer and help each other to fully understand how and when to use the substitution method.

Example 2

Q Solve the following linear system using the substitution method.

$4x - 3y = -11, -5x + 2y = 12$

A
$$-5x + 2y = 12$$
$$+5x \qquad +5x$$

          Isolate one of the variables in one of the equations.

$$2y = 5x + 12$$

$$\frac{2y}{2} = \frac{5x}{2} + \frac{12}{2}$$

$$y = \frac{5}{2}x + 6$$

▶ Continued on next page

NUMERICAL AND ABSTRACT REASONING

$$4x - 3\left(\frac{5}{2}x + 6\right) = -11$$

Substitute the expression into the other equation to solve for the other variable.

$$4x - \frac{15}{2}x - 18 = -11$$

Use the distributive property to expand the brackets.

$$2\left(4x - \frac{15}{2}x - 18 = -11\right)$$

Multiply each side of the equation by 2 to eliminate all of the denominators.

$$8x - 15x - 36 = -22$$
$$-7x - 36 = -22$$
$$+36 \quad +36$$

Solve for x.

$$-7x = 14$$

$$x = -2$$

$$y = \frac{5}{2}(-2) + 6$$

Substitute the value of x into the equation for y to find the value of the other variable.

$$y = 1$$

$$4(-2) - 3(1) = -11$$
$$-11 = -11 \quad \checkmark$$

Verify your answer in each of the original equations.

$$-5(-2) + 2(1) = 12$$
$$12 = 12 \quad \checkmark$$

The solution is $(-2, 1)$.

Write the solution as an ordered pair.

Activity 4 – The Paradigm Project

According to the World Health Organization, approximately 4 million women and children die every year from lower respiratory diseases related to indoor cooking smoke. Globally, this is the number one cause of death in children under 5 years of age. The Paradigm Project is a social entrepreneurial organization that is trying to address this serious issue by supplying clean-burning cooking stoves to people in developing countries.

A family living in a poor rural area in Guatemala has an income of $5 per day. They use 30% of their income to pay for cooking fuel each day. If they buy a clean-burning stove costing $35, they will only use 15% of their income on cooking fuel. How many days will it take for the stove to pay for itself with the money they save on cooking fuel?

1 Copy and complete this table.

Number of days	Total cost of fuel (no stove)	Total cost of fuel (including new stove)
0	0	$35
1		
2		
3		
4		
5		

2 Graph each relationship and solve the linear system.

3 Determine the equation of each line, and hence create a system of linear equations.

4 Verify your answer by solving the system by substitution.

5 What are the other benefits of the clean-burning stove, apart from the fuel efficiency savings?

6 Once the stove is paid off, what else do you think the family will buy with the money it saves on fuel?

7 How will the stove improve the overall quality of life for the family?

Practice 4

1 Solve the following linear systems using substitution. Verify each answer.

a $4x + y = 1$, $2x + 3y = -7$ **b** $x - 7y = 0$, $3x - 4y = 17$

c $5x - y = -1$, $-10x + 3y = 4$ **d** $2x - 5y = 6$, $-3x + 4y = -6$

e $5x - 2y = 6$, $x = 3y - 4$ **f** $x = -5$, $4x + 2y = -18$

2 Solve the following systems by graphing and by substitution. Verify each answer.

a $y = 6x - 5$, $y = 3 - 2x$ **b** $4x - y = -8$, $2x + 3y = 10$

c $5x - 3y = -2$, $-3x + 7y = 9$ **d** $3x + 7y - 1 = 0$, $2y = 8 + 3x$

e $y = -2$, $-5x + 6y = -2$ **f** $6x + 4y = 14$, $5x - 2y = 9$

3 Imagine you are starting a business that sells cell phone covers with the aim of donating money to a local charity. Your start-up costs for the business are $1300 and each cell phone cover costs $30 to manufacture. You sell the phone covers for $60 each.

  a Explain why the equations $y = 60x$ and $y = 30x + 1300$ represent the revenue and cost functions.

  b Solve the system by substitution to find how many covers you must sell in order to break even.

  c Find how much revenue you will make at the break-even point.

  d When will you start to make a profit? Explain.

4 Which method do you prefer, graphing or substitution, or does it depend on the question? Explain.

Solving linear systems by elimination

As you have seen, solving a linear system by substitution can be very efficient, especially when it is fairly straightforward to isolate one of the variables. The method of elimination is another method that can be used to solve any linear system. This method does not rely on isolating a variable.

Activity 5 – Equivalent equations

1 Given the equation $4x - 2y = 8$,

   a multiply the equation by 2

   b multiply the equation by -3

   c divide the equation by -2.

   d Explain why the equations you created in **a** through **c** are equivalent to the original equation.

2 Given the equation $-4x + 5y = -7$,

   a multiply the equation by an integer so that the coefficient of x is 12

   b multiply the equation by an integer so that the coefficient of y is -25

   c multiply the equation by an integer so that the coefficient of x is 4.

   d By what number could you multiply the equation so that the new equation is **not** equivalent to the original? Explain.

The elimination method uses the concept of equivalence in order to solve a linear system. Creating equivalent equations allows you to manipulate them more easily.

ATL
1 & 2

Pairs

Activity 6 – Elimination method

Perform the following activity in pairs.

Your school is putting on an event to support an upcoming fundraising walk. Visitors to the event can buy the 2/1 combo (2 slices of pizza and a drink) for $5.20. They can also buy the 6/2 combo (6 slices of pizza and 2 drinks) for $14.40. How much does one slice of pizza cost? How about one drink?

1 Starting with the more expensive combo, what happens if you subtract the first combo from it? How many slices of pizza and how many drinks are left and what is their cost? Explain your answer.

2 Repeat the same process, subtracting the 2/1 combo from your result in step 1.

▶ Continued on next page

3 What does this tell you? What is the price of a slice of pizza? What is the price of a drink?

4 Represent the original combos with a system of equations. Use the variable p to represent the number of slices of pizza and d to represent the number of drinks.

5 Multiply the combo 1 equation by 2 and then subtract this equivalent equation from the combo 2 equation.

6 How does this relate to what you did in steps 1 and 2? Why do you think this method is called elimination?

7 Refer back to the linear system you created in step 4. Multiply the first equation by −3 and then add it to the second equation.

8 What are the similarities and differences between step 5 and step 8? Explain.

9 Find the value of d. Verify your solution in both original equations.

Consider the linear system $2x - 5y = 12$
$$2x + 20y = -18$$

10 How could you combine the two equations to eliminate the variable x? Explain.

11 How could you combine the two equations to eliminate the variable y? Explain.

12 Solve the system by elimination. Verify your answer.

13 Summarize the process of solving a linear system by elimination.

Reflect and discuss 5

- You can either add or subtract equations when you use the elimination method. How does your choice of operation affect the value by which you multiply the equation(s)? Explain.

- Which operation will result in making fewer mistakes, adding or subtracting? Why?

- Once you have the value of the first variable, you need to find the value of the other one. What options do you have for finding the value of the second variable?

Example 3

Solve the following linear system using the elimination method.

$$2x - 3y = 1$$

$$3x + 5y = \frac{3}{2}$$

$$(2x - 3y = 1) \times 3$$

$$\left(3x + 5y = \frac{3}{2}\right) \times -2$$

In order to eliminate a variable, **both** equations need to be multiplied by a value. Multiply the top equation by 3 and the bottom equation by -2 so that the coefficients of x are 6 and -6, and then add the equations.

$$6x - 9y = 3$$
$$\underline{-6x - 10y = -3}$$
$$-19y = 0$$

$$2x - 3(0) = 1$$

$$2x = 1$$

$$x = \frac{1}{2}$$

Substitute the solution into one of the equations to solve for the other variable.
Another option is to use elimination again, but this time eliminate the y.

$$2\left(\frac{1}{2}\right) - 3(0) = 1$$

$$1 - 0 = 1 \quad \checkmark$$

Verify your answer in each of the original equations.

$$3\left(\frac{1}{2}\right) + 5(0) = \frac{3}{2}$$

$$\frac{3}{2} = \frac{3}{2} \quad \checkmark$$

The solution is $\left(\frac{1}{2}, 0\right)$.

Write the solution as an ordered pair.

Activity 7 – Benefit concerts

A benefit concert is a musical performance in which musicians use their celebrity status to help raise money and awareness for a particular humanitarian crisis, either local or abroad. Some large and famous concerts have been Live Aid (1985), Conspiracy of Hope Tour (1986), The Concert for New York City (2001), The SARS Benefit Concert (2003), Live 8 (2005), Live Earth (2007), One Love Manchester (2017), and Hand in Hand: A Benefit for Hurricane Relief (2017).

1 Research these big concerts to see what they were raising money and/or awareness for.

Imagine that your school has an ongoing connection with an orphanage. You would like to organize a concert to raise money to help build a school for the children in the orphanage. You have managed to get some great local bands to play and they have agreed to do the concert for 20% of the price of each ticket sold. The rental of the auditorium and crew is $4000. You are going to sell tickets to the concert for $20.

2 Write down an equation to represent the **revenue** from the concert in terms of the number of tickets sold. Justify your equation.

3 Write down an equation to represent the **cost** of putting on the concert in terms of the number of tickets sold. Justify your equation.

4 Solve the revenue–cost system by graphing, and find the break-even point.

5 Solve the system by elimination.

6 The auditorium has a capacity of 1000 people and your goal is to raise a total of $10 000. Is this going to be possible? If so, how many tickets will you need to sell? Show your working.

Investigation 3 – Classifying systems of equations II

Here are three linear systems.

A	B	C
$y = 2x + 1$	$2y + 8 = x$	$6x = 20 - 2y$
$y = -x + 7$	$4y - 2x = -16$	$y + 3x - 5 = 0$

1 You have already graphed these linear systems in the first investigation on classifying systems (Investigation 2). Write down the classification of each system based on those results.

2 Solve each system by substitution.

3 Solve each system by elimination.

4 Write a brief summary of the results of your investigation into how to classify a linear system when using an algebraic method.

5 Verify your results with at least two more linear systems.

6 Justify your results. Why do they make sense?

Practice 5

1 Solve each system of linear equations using the method of elimination.

a $2x - 5y = 9$
   $3x + 2y = 4$

b $5x - 3y = 5$
   $3x + 2y = 4$

c $x + 5y = -2$
   $-3x - 6 = 15y$

d $6x + 9y = 12$
   $-4x - 6y = -8$

e $3x - y - 11 = 0$
   $x + 2y = 6$

f $x - 3y = 5$
   $3x + 2y = 4$

g $8x + 5y = 2$
   $5x + 2y = 8$

h $4x - 8y = -3$
   $-10x + 20y = 12$

i $3x - y = 11$
   $x + 2y = 6$

j $\dfrac{x}{2} - \dfrac{y}{3} = 1$
   $x - \dfrac{y}{4} = 2$

▶ Continued on next page

2 a Verify three of the linear systems in question 1 by graphing.

b How did you decide which linear systems to verify using the method of graphing? Explain.

3 a Verify three of the systems in question 1 using the method of substitution.

b How did you decide which linear systems to verify using the method of substitution? Explain.

Reflect and discuss 6

ATL
1 & 2

Pairs

Discuss the following with a peer.

- Explain when you would use the graphing method to solve a linear system.

- Explain when you would use substitution to solve a linear system.

- Explain when you would use elimination to solve a linear system.

- Make a flow chart of the different strategies you would use depending on the information given. Be sure to include which graphing method to use when graphing a linear system (y-intercept and gradient, find x- and y-intercepts, find two points).

Problem solving with linear systems

When decisions need to be made based on a variety of factors, they can often be modeled using linear relationships. Knowing how to solve a linear system can help you to make an informed choice.

Activity 8 – Creating a system of equations

Translating information into a system of equations is an important first step in solving a problem. With a partner, copy and fill in this table.

Information	Defining variables	Creating equations	Solving the system. Which method? Justify your choice.
The sum of two numbers is 1211 and their difference is 283.	Let S represent the smaller number. Let L represent the larger number.	$S + L = 1211$ $L - S = 283$	
Four times the mass of a baseball is 16 g less than the mass of a basketball. The sum of their masses is 756 g.	Let b represent ____ Let B represent ____		
Two hamburgers and a drink cost $5.00. Three hamburgers and two drinks cost $7.90.	Let __ represent ____ Let __ represent ____		
A rectangle with a perimeter of 28 m has a length which is 2 m less than triple its width.	Let __ represent ____ Let __ represent ____		
The Tritons made a total of 42 shots in a basketball game. They scored a total of 96 points through a combination of 2-point and 3-point shots.	Let __ represent ____ Let __ represent ____		
Seven times the smaller of two numbers plus nine times the larger number is 178. When ten times the larger number is increased by eleven times the smaller number, the result is 230.	Let __ represent ____ Let __ represent ____		

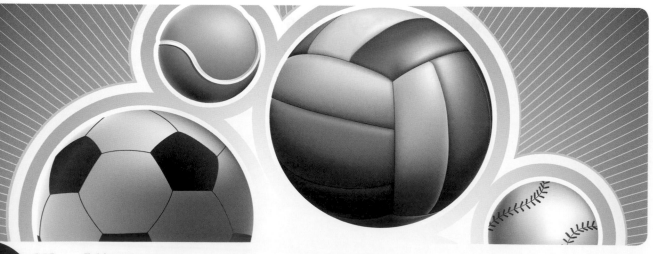

Organizing the information in problems is sometimes made easier with the use of a table.

Example 4

Q You receive $325 in donations, some in $5 bills and others in $10 bills. If you counted 46 bills in all, how many of each type of bill do you have?

A Let x represent the number of $5 bills.
Let y represent the number of $10 bills.

> Write a statement explaining what each of the variables represents.

	Number	Value ($)
$5 bills	x	$5x$
$10 bills	y	$10y$
TOTAL	46	325

> Setting up a chart might help organize the information – each column contains information to set up an equation.

$$x + y = 46$$
$$5x + 10y = 325$$

> Set up two equations to represent the information in the question.

$$x + y = 46$$
$$y = 46 - x$$

> Select the most appropriate and efficient method to solve the question – either substitution or elimination. In this case use substitution.

$$5x + 10y = 325$$
$$5x + 10(46 - x) = 325$$
$$5x + 460 - 10x = 325$$
$$-5x + 460 = 325$$
$$-5x = -135$$
$$x = 27$$

> Solve.

$$x + y = 46$$
$$27 + y = 46$$
$$y = 19$$

> Substitute the solution into one of the original equations to solve for the other variable.

$$27 + 19 = 46 \quad \checkmark$$
$$5(27) + 10(19) = 325 \quad \checkmark$$

> Verify your answer in the original equations.

> Write a concluding sentence with the answers.

You have 19 $10 bills and 27 $5 bills.

Example 5

Q You invested $12 000 into two types of bonds. One bond yields 8% interest each year. The other bond yields 12% interest each year but has more risk. If you earned $1160 interest in a year, how much did you invest in each bond?

A Let x be the amount invested at 8%.
Let y be the amount invested at 12%.

> Write a statement explaining what each of the variables represents.

Note: risk is not factored into the question at all.

	Amount invested	Interest rate	Interest earned
8%	x	0.08	$0.08x$
12%	y	0.12	$0.12y$
Total	12 000		$1160

> Setting up a chart might help organize the information – each column contains information to set up an equation.

$$x + y = 12\,000$$
$$0.08x + 0.12y = 1160$$

> Set up two equations to represent the information in the question.

$$-0.08x - 0.08y = -960$$
$$+0.08x + 0.12y = 1160$$

> Select the most appropriate and efficient method to solve the question – in this case either substitution or elimination. Use elimination.

$$0.04y = 200$$
$$y = 5000$$

> Multiply the first equation by –0.08

> Solve.

$$x + y = 12\,000$$
$$x + 5000 = 12\,000$$
$$x = 7000$$

> Substitute the solution into one of the equations to solve for the other variable.

$$7000 + 5000 = 12\,000 \quad \checkmark$$

> Verify your answer in the original equations.

$$0.08(7000) + 0.12(5000) = 1160$$
$$560 + 600 = 1160 \quad \checkmark$$

You invested $5000 in the higher risk bond (12%) and $7000 in the lower risk bond (8%).

> Write a concluding sentence with the answers.

Reflect and discuss 7

Can you think of a real-life situation that could be modeled using a linear system in which there is no solution? Can you think of one in which there are an infinite number of solutions?

Activity 9 – Go big?

A coffee farmer in Costa Rica currently has costs of 80 cents per kilogram to grow and harvest the coffee, plus $1000 of fixed costs for land management per year. She can sell her green, unroasted beans for a guaranteed $1.50 per kilogram.

1 Create a linear system model for this scenario.

2 Solve the system using two of the three methods (graphing, substitution, elimination) in order to calculate the break-even point.

The farmer knows that her coffee is being sold to consumers for $8 per kilogram and is considering getting into the rest of the supply side of the business. A social entrepreneur said she would help her start up a farmers' cooperative where they will grow, roast, package and export the coffee directly to the customers. The costs associated with the entire process are $5 per kilogram and the fixed costs for growing, processing and exporting the coffee would be substantially more at $8000 per year.

3 Create a linear system model for this new scenario.

4 Solve the system using the method you did not use before in order to calculate the break-even point.

5 Verify your answer using one of the other two methods.

6 Determine the profit function for both scenarios. The farmer produces 6000 kilograms of coffee per year. What is her profit in both scenarios?

7 Based on your analysis, what would you recommend that she do? Justify your answer.

▶ Continued on next page

8 Why is there such a large difference between the profit made in the two scenarios?

9 What are the advantages and disadvantages of starting the cooperative?

Practice 6

ATL2 Answer the first three questions in pairs. When someone encounters difficulty, ask leading questions instead of giving the answers.

Pairs

1 Ann has $300 made up of $5 and $10 bills. If there are 39 bills in all, how many $5 bills does she have?

2 A parking machine contained $3.05 made up of dimes and quarters. There were 20 coins in all. How many dimes were there?

> A dime is 10 cents and a quarter is 25 cents.

3 Saira invested $1000, part at 8% per annum and the remainder at 9% per annum. After one year, her total interest from these investments was $84. How much did she invest at each rate?

4 Plan International offers gifts of hope options where you select a gift to support a community in a developing country. It costs €500 to equip a schoolroom and €295 to send a girl to school. If you buy 20 gifts at a total cost of €6515, how many classrooms will be equipped and how many girls will be sent to school?

5 A professional soccer player is negotiating her contract. Using the advice of her manager, she asked for $800 000 for the year, plus an additional $1500 for every game she plays in. The team offered $6000 for every game played and $600 000 for the year. How many games would she need to play for the team's offer to be the better option?

▶ Continued on next page

6 Supplying a sheep to a family costs AU$50 and a bee hive costs AU$35. If there were 24 gifts totaling AU$990, how many of each were bought? What would be the benefit of supplying a sheep or bee hive to a family?

7 Jen invested $8000, part at 9% and the rest at 10%. The interest after one year was $740. How much did she invest at each rate?

8 Two social events have been organized by a local scout group to raise funds for a well in Tanzania. During the first night, 25 children and 20 adults attended and the revenue for that evening was £150.00. On the second night there were 30 children and 22 adults, with revenues of £170.00. How much did each adult pay to attend this social event, assuming that the ticket prices were the same for both events?

9 Many companies today are moving toward more ethical supply chains. Costco Wholesale Corporation has a code of conduct for its global suppliers which prohibits human rights abuses in their supply chain, for example human trafficking, physical abuse of workers and unsafe working environments. To ensure the code of conduct is being enforced, they may audit the facilities of suppliers, especially those in countries that are more at risk of such violations. A basic Costco membership in the United States is $60 per year, while the premium membership is $120 per year. You get 2% back on Costco purchases with the premium membership and nothing back with the basic membership. How much money would you have to spend at Costco in a year before the premium membership is the better option for you to buy? Set up a linear system to show the membership options and use it to solve the problem.

10 A contractor employed 5 adults and 3 teenagers for one day and paid them a total of $224 per hour. The following day she employed 3 adults and 5 teenagers for a total of $160 per hour. What was the hourly rate paid to each adult and teenager? (Assume that all the adults are paid the same rate and all the teenagers are paid the same rate.)

11 Over 1 billion people around the world live where the electrical grid service is unreliable and at least 1 billion more people live completely out of range of the grid. A social entrepreneur manufactures solar lanterns for which the cost of materials, in Indian rupees, is Rs 350 per lantern and the fixed costs are Rs 800 000 per month.

a If each lantern sells for Rs 475 (costs are kept low to ensure that the people who really need them can buy them), find how many lanterns must be made and sold each month for the company to break even.

b What is the revenue at the break-even point? Convert this amount into your local currency.

▶ Continued on next page

12 Find the equation of the line, in standard form, that passes through $(-7, 3)$ and the point of intersection of $4x - y = 3$ and $2x + y = 9$.

13 A rectangle has vertices at $A(2, 2)$, $B(2, 6)$, $C(4, 6)$ and $D(4, 2)$. Develop an **algebraic** solution to determine the coordinates of the point of intersection of the diagonals.

14 Thailand is one of the world's top rice producers. For a typical rice farm, the fixed costs (in Thai baht) are 16 000 THB per growing season and the variable costs are 12 000 THB per hectare. Rice is sold for 19 500 THB per hectare.

a Set up a linear model to find how many hectares you would need to break even.

b If the average size of a rice farm is 6 hectares, what profit does the farmer make?

c Convert this amount to your local currency. Do you think this is enough money to live on for the whole growing season? Explain.

d Suppose the price of rice is trending downward. The selling price of rice is set by the global market and the local small farmer has no control over fluctuations in prices, which can be significant. If the selling price drops by 15%, as it has recently, what will be the new break-even point? How will this affect Thai rice farmers?

Formative assessment – Sustainable denim

criteria
C, D

More companies are starting corporate social responsibility (CSR) reporting, sometimes referred to as the *triple bottom line* (triple for "people, planet and profit"). This means that the companies take the initiative to ensure that their supply chains are ethically and environmentally responsible and that they have positive interactions with the communities in which they operate.

Unfortunately, profit is still the driving force of many companies and in order to cut costs, employees, especially those in developing countries, may work in dangerous and unhealthy conditions. Pollution in manufacturing and shipping can be extensive and the Earth's natural resources can be used in an unsustainable way.

▶ Continued on next page

Case study:

Two premium denim jean factories in Asia have the following costs per month. The first one is a socially responsible factory and the second is a traditional factory of a type still common in Asia.

Cost	Socially responsible factory	Traditional factory
Materials per pair of jeans	$12*	$7
Wash/dye process per pair of jeans	$2**	$0.25
Labour cost per pair of jeans	$1.50***	$0.50
Rental of building, utilities, admin costs, programs, etc.	$10 000****	$5800

* All raw materials that go into making the denim jeans meet the highest ethical and environmental standards.

**The factory uses organic cotton and a water recycling system to reduce the environmental impact and not contaminate local community water supplies. It also uses non-toxic organic vegetable dye to ensure that no harm is done to the workers, rather than the harmful chemical dyes used in traditional jean manufacturing.

*** Workers earn a living wage, which is 3 times the average of workers in the industry, allowing them to better support themselves and their families and even to save for the future.

**** Fixed costs are higher due to better ventilation, light, space and general facilities in the factory. English language courses and training courses in all components of jean production require more time and every worker is rotated to ensure they are skilled in every component of jean making.

Build a linear systems model

a Derive a cost function for both factories.

b If the jeans are sold to the wholesaler for $30 a pair, determine a revenue function.

c Graph the cost functions and the revenue functions of both factories on one set of axes.

d What is the break-even point for both factories? Comment on the difference between the two.

e Verify your answer by solving the system algebraically. Show your working.

f At what price would the socially responsible factory have to sell a pair of jeans if they wanted the same break-even point as the traditional factory? Round your answer appropriately and explain the degree of accuracy.

g How much more is this per pair as a percentage of the original price?

▶ Continued on next page

Reflect and discuss 8

- Describe whether or not your solution makes sense in the context of the problem.

- Given the steps still needed to get these jeans to the end customer (including shipping, marketing and product placement in stores), the final selling price of the jeans could be up to 5 times the amount the factory sells them for. What would the approximate final price be of both pairs of jeans?

- Do you think companies should put sustainable supply chains in place, even though this decreases their profit? Explain.

ATL1 Discuss the following questions in groups of 3.

- Whose responsibility is it to ensure sustainable growth? Companies? Governments? Consumers?

- Each person in your group takes the role of one of these key players and lists how they could help generate change. Who do you think is in the best position to generate change?

- How could a factory with ethical practices such as the one described in the case study affect current and future generations?

- If you were to write a headline that captured the most important aspect of this task and your discussion, what would that headline be? Share your headlines with the class.

Research "social entrepreneur premium denim" to see what some companies are doing in the denim industry and broader textile industry. Find out the motivation for creating these companies.

Activity 10 – Urban planning

A 10-acre plot of forest has been designated for a housing development due to the expansion of a nearby city. One proposal comes from a new company that aims to create a CO_2 neutral community. In other words, they want to develop the housing estate so that enough trees are left on the land to convert the CO_2 of the people living in the estate. The problem they must solve is determining the number of houses they can build.

Factors to consider:

An average human exhales around 2.3 pounds of CO_2 in a day.

The average tree can convert 48 pounds of CO_2 per year.

There are currently approximately 700 trees per acre in the forested region.

Each house will have a 3000-square-foot plot.

▶ Continued on next page

Process:

1 What units will you use? Explain.

2 Given that x will be the number of houses and y will be CO_2 production per year, what does the gradient of each line represent?

3 Assuming an average of 4 people living in each house, determine the linear equation that represents the people living in the housing development. Make sure you show all of your working to calculate the gradient.

4 Is the gradient positive or negative? Explain.

5 What does the y-intercept represent?

6 Determine the linear equation that represents the trees living in the housing development. Make sure you show all of your working to calculate the gradient.

7 Will the gradient of this line be positive or negative? Explain.

8 What does the y-intercept represent in this scenario?

9 Graph both linear equations on the same set of axes, making sure that you clearly show the point of intersection. This can be done using technology (graphic calculator or graphing software). If using technology, make sure that you take screen shots of your solutions and copy them into your report, which you will need to hand in.

10 What does the point of intersection represent?

11 Verify your solution using an algebraic method. Show all working.

12 Given the number of houses that can be built and still remain ecologically viable, calculate the percentage of the original plot of land that will be used for housing and the percentage that will be left forested.

13 When housing developments are being built, are these percentages maintained? How are they different? Why do you think that is?

Changing the model:

14 What would happen if you used 5 people living in each house as the average? Which line would change? How would it change? Determine the new linear equation and compare it with the originals. How would that affect the point of intersection?

15 What would happen if there were only 500 trees per acre of the forest? Which line would change? How would it change? Determine the new linear equation and compare it with the originals. How would that affect the point of intersection?

Unit summary

A *linear equation* is an equation of degree one.

A *quadratic equation* is an equation of degree two.

A *cubic equation* is an equation of degree three.

An equation generally has the same number of solutions as its degree.

A system of equations is a group of equations that can be solved for each of the variables it uses. With two variables, you will need at least two equations in order to find the values of both variables.

If the two equations are both linear, you have what is called a system of linear equations or simply a *linear system*.

Linear systems can be solved by graphing each of the lines and finding the point of intersection.

Linear systems can also be solved algebraically using substitution or elimination.

To solve a linear system by substitution, rewrite one of the equations to isolate one of the variables. Then, substitute this expression into the other equation. Solve this new equation.

To solve a linear system by elimination, multiply one or both equations so that the coefficients of x or y are exact opposites and then add the equations. This will eliminate the variable and then you can solve for the other by substitution.

Always verify your answers by substituting into the original equations.

Linear systems can have no solution, one solution or infinitely many solutions.

Graphically, a linear system has no solution when the lines are parallel to each other. There are infinitely many solutions when you are given the same line twice.

Algebraically, there are infinitely many solutions when all of the variables are eliminated and you end up with a true equation, for example $0 = 0$ or $-4 = -4$.

Algebraically, there is no solution when all of the variables are eliminated and you end up with an equation that is not true, for example $0 = 2$ or $-4 = 7$.

Unit review

📖 **Launch additional digital resources for this chapter**

Key to Unit review question levels:

| Level 1–2 | Level 3–4 | Level 5–6 | Level 7–8 |

1 **Solve** each equation and **verify** your answers.

   a $2x + 7 = 11$

   b $3(5 + 2x) = -21$

   c $5(x + 4) = 3(x - 3) + 5$

   d $\dfrac{x+3}{4} = \dfrac{x-1}{2}$

   e $\dfrac{1}{2}y - 3 = \dfrac{2}{3}y + 4$

   f $\dfrac{3}{4}(y + 2) = 2y - 11$

2 **Sketch** the following graphs and **write down** the coordinates of each point of intersection.

   a $y = -3x + 10$ and $y = 2x - 5$

   b $y = 2x - 6$ and $y = \dfrac{3}{4}x - 6$

3 Find the **exact** intersection point of the following two linear systems using technology.

   a $y = x - 8$ and $y = -\dfrac{1}{3}x - 4$

   b $x = 5y - 75$ and $x + \dfrac{1}{2}y - 7 = 0$

4 **Solve** the following linear systems algebraically.

   a $y = 6x + 7$ and $3y + 12x - 6 = 0$

   b $3a + 2b = 16$ and $2a + 3b = 14$

   c $14x + 21 = -21y$ and $2x + 3y = 12$

   d $2y - 4 = x$ and $3y - 6x = -3$

   e $6x = 2y - 8$ and $5y - 5x + 10 = 0$

5 Classify each of the following linear systems and **justify** your answers.

  a $4x - 3y = 15$ and $8x - 9y = 15$

  b $4x - 3y = 5$ and $8x - 6y = 10$

  c $2x = 3y - 6$ and $4x - 6y = 24$

6 When the sum of four times a number and ten is divided by five, the result is negative 2. Find the number.

7 Two consecutive integers are both smaller than 20. If three times the smaller integer is equal to eight less than double the larger integer, find both integers.

8 A four-sided figure with a perimeter of 120 cm has a width which is 30 cm less than twice its length. What kind of quadrilateral is this? **Justify** your answer.

9 The difference between two numbers is divided by 2 and the result is 3. Find these two numbers if triple the smaller number is 4 less than double the larger number.

10 You have three dollars' worth of coins in your piggy bank consisting of nickels, dimes and quarters. The number of dimes is double the number of nickels. The number of quarters is one-third of the number of nickels. How many coins are in your piggy bank in total?

> A nickel is 5 cents, a dime is 10 cents and a quarter is 25 cents.

11 You could use a music streaming service and pay $9.99 a month or pay an average of $1.14 for each song you download.

  a Graph this scenario using technology.

  b How many songs a month would you have to download to make the streaming service the cheaper option?

  c Can you think of why you would choose the pay-per-song option instead, even if you downloaded 20 songs a month?

12 Some teachers are taking a group of Year 3 students to a talk on sustainable supply chains at which some social entrepreneurs will speak. 42 tickets cost $582 in total. If each student's ticket costs $12 and each teacher's ticket costs $25, use an algebraic method to find the number of teachers and students going to the talk.

13 A microlender invested AU$20 000 in a combination of two ventures, one with low risk yielding 3% per annum and one with higher risk yielding 8% per annum. If the interest after one year was AU$850, **calculate** algebraically the amount invested in each venture.

14 In some communities in the United States, local food trucks have started giving out meals to the homeless. While the trucks still have delicious meals for sale, people who cannot afford them can get a meal for free. The cost to make each meal is $2.50 and it is sold for $8. The food truck costs $3000 per month to operate. They must make a profit of at least $2000 each month to cover the costs of the free meals. What is the least number of meals they need to make and sell in a month (assume 30 days) to cover the cost of the free meals?

15 In some areas of the world, you may be living on land where you could drill for your own gas. A farmer pays $5000 for natural gas each year. If you drill yourself, it costs $40 000 to start up your own gas line and you could drill enough to produce $7500 worth of gas each year and then sell what you don't use into the grid. How many years would it take to pay off the gas line costs? Would this be a good investment? What other factors would you have to consider?

16 You are boxing up care packages to send to developing countries. The first type contains bed nets to prevent mosquito bites and the spread of malaria (€10 each) and solar power kits (€60 each) to provide light in areas with restricted or no electricity. The second contains medicine kits for mothers and babies (€20 each) and baby blankets and supplies (€15 each). There are 140 items in the first package. In the second package, the number of baby blankets is twice the number of medicine kits.

If the package with the medicine kits and blankets cost €3000 and the package with the nets and solar kits cost €3400, **calculate** how many of each of the four items there are in the packages.

Summative assessment

Running your own business for a day

criteria C, D

Part 1 – A fundraising event

Your school will be hosting a fundraiser, at which students will set up booths at a market to sell products that they have made. Research and find a charity or organization that you would like to support by raising awareness and donating all your profits from the event. Then create a one-page flyer summarizing the charity to show to customers and hand out during the event.

Part 2 – Business costs and pricing

You will need to set up a business to make your product.

a Decide what you would like to make (for example, baked goods, a simple toy, jewellery).

b Go online to find the approximate cost of everything you will need to start up your business (including buying all the tools or utensils needed to make your goods). Make sure you itemize all of the costs so that you can include them in your report.

c Approximate the cost of making each individual item to be sold (i.e. of the materials you will use to make the product). Make sure you show all of your working to calculate the cost per item and include it in your report.

d Decide how much you would like to charge for your product. It must be a realistic selling price so that you will be able to sell the product but also raise money for your charity.

Part 3 – Create a linear model

a Determine the equation that represents the cost of making the product. What do the gradient and y-intercept represent in this equation?

b Determine the equation that represents revenue (the money that you make by selling the product). What do the gradient and y-intercept represent in this equation?

c Graph the linear system that you have created. Remember to label the axes as well as the point of intersection.

d The point of intersection represents the break-even point, where costs equal revenue, so the profit at this point is zero. How many of your product do you have to sell to break even? (Solve this graphically.)

e Verify your point of intersection by solving the system algebraically.

f Write a general statement relating the quantity of your product sold to when you are losing money and when you are making a profit.

g How much money will you lose or make if you make and sell

  i 10 units of your product?

  ii 1000 units of your product?

h What happens to the original line (gradient and y-intercept) and the point of intersection (number of products you need to sell in order to break even) when you alter each of the following parts of your linear system?

  i Change of start-up costs

  ii Change of variable costs (material costs)

  iii Change of selling price

Show each of the above three scenarios in a sketch on three separate sets of axes – drawing the original cost and revenue lines in one colour and the new lines (showing both an increase and a decrease) in different colours. Label the original point of intersection and the new break-even points.

> This is conceptual so you do not need to calculate the actual break-even points – just show how the lines have shifted and how the point of intersection has changed.

Part 4 – Your report

Use publishing software to create a professional-looking, eye-catching flyer that summarizes the information about the charity/organization.

In your report include:

- a list of costs, with an explanation and calculations showing how you derived all the costs and how you determined the selling price
- a linear model which graphically shows cost and revenue functions and break-even point
- verification of break-even point using algebra
- three sketches showing how changes in costs and selling price alter the graphs and break-even points.

Reflect and discuss

- Explain the degree of accuracy of your solution.
- Describe whether or not the answer makes sense in the context of the problem.
- Given that you will only have one day to sell your product, how many units of your product do you anticipate selling? Using the point of intersection that you calculated, will you make or lose money using your estimated selling quantity?
- What could you do to ensure that you make money? Can you even make such assumptions?
- What other factors do you need to take into consideration when determining how many of your product you will make for the school fundraiser?
- Given the projected break-even point, is there anything about your product/model that you would change? If so, what would it be?
- How can you make a difference as a social entrepreneur?
- What is our responsibility to those in our community and other communities?

Answers

1 Number

Practice 1

1 a Rational   **b** Rational   **c** Rational   **d** Irrational   **e** Rational
   f Rational   **g** Irrational   **h** Rational   **i** Irrational   **j** Rational

2 a $\dfrac{19}{50}$ **b** $\dfrac{7}{33}$ **c** $\dfrac{8}{5}$ **d** $\dfrac{22}{9}$ **e** $\dfrac{6437}{500}$

   f $\dfrac{2}{15}$ **g** 5 **h** $-\dfrac{2587}{625}$ **l** $\dfrac{312}{55}$ **j** $\dfrac{2033}{225}$

3 a $\dfrac{1}{3}, \dfrac{2}{3}$ **b** Individual response   **c** $\dfrac{25}{99}, \dfrac{74}{99}$

   d Individual response. Possible answers are of form $0.\overline{abc}, 0.\overline{xyz}$ where abc, xyz are three digit numbers which add up to 999.

Practice 2

1 a $\dfrac{1}{36}$ **b** 1 **c** $\dfrac{1}{1000}$ **d** 1 **e** 64

   f $-\dfrac{1}{9}$ **g** 343 **h** $\dfrac{1}{64}$ **l** 1 **j** $\dfrac{1}{64}$

   k $-\dfrac{1}{27}$ **l** $\dfrac{16}{9}$ **m** 1 **n** $\dfrac{7}{4}$ **o** $\dfrac{125}{8}$

   p $-\dfrac{1}{144}$ **q** $\dfrac{1}{144}$ **r** 0 **s** $\dfrac{9}{4}$ **t** $\dfrac{1}{32}$

2 a 4^6 **b** 5^{-5} **c** 11^{-2} **d** 10^{-4} **e** $(-7)^5$

3 a $4^{-2}, 5^0, 3^1, \left(\dfrac{1}{3}\right)^{-2}$ **b** $2^{-2}, 3^{-1}, \left(\dfrac{5}{8}\right)^0, \left(\dfrac{1}{4}\right)^{-1}$ **c** $7^{-1}, \left(\dfrac{2}{5}\right)^2, 2^{-2}, 1^{-3}$ **d** $0^1, \left(\dfrac{1}{2}\right)^2, 3^{-1}, 1^0$

4 "Zero 5s" is multiplying, not taking an exponent

5 Correct.

6 "Half of an 8" is multiplying, not taking an exponent

7 Correct. A positive exponent is the number of times the base value is multiplied and a negative exponent is the number of times 1 is divided by the base.

8

Prefix	Exponential form	Expanded form
giga	10^9	1 000 000 000
mega	10^6	1 000 000
kilo	10^3	1000
deci	10^{-1}	$\dfrac{1}{10}$
centi	10^{-2}	$\dfrac{1}{100}$
milli	10^{-3}	$\dfrac{1}{1000}$
micro	10^{-6}	$\dfrac{1}{1\,000\,000}$
nano	10^{-9}	$\dfrac{1}{1\,000\,000\,000}$
pico	10^{-12}	$\dfrac{1}{1\,000\,000\,000\,000}$
femto	10^{-15}	$\dfrac{1}{1\,000\,000\,000\,000\,000}$
atto	10^{-18}	$\dfrac{1}{1\,000\,000\,000\,000\,000\,000}$

Practice 3

1
a 5^9
b $\dfrac{1}{6^4}$
c $\dfrac{1}{8^4}$
d x^5

e y^5
f s^{11}
g $\dfrac{1}{10^7}$
h $\dfrac{1}{12^{12}}$

i 9^2
j $\dfrac{1}{2^2}$
k $\dfrac{1}{x^{10}}$
l a^5

m $\dfrac{1}{w^{5q}}$
n $\dfrac{1}{m^{2t}}$
o $\dfrac{1}{y^{11t}}$
p $\dfrac{1}{a^6}$

q $\dfrac{30}{x^7}$
r $6a^3b^2c^5$
s $-14p^4q^4$
t $\dfrac{3}{y^4z^2}$

u $\dfrac{7qr^3}{p}$
v $\dfrac{ef}{3}$
w $\dfrac{q^3}{2p^6}$
x $\dfrac{5}{a^5b^3}$

2
a 10^{15}
b 10^{18}
c 10^{12}
d $\dfrac{1}{10^{24}}$
e $\dfrac{1}{10^5}$

3
a 3^2
b 3^4
c 10^{-1}
d equal
e 2^{-5}
f equal

4
a $\dfrac{1}{2}$
b 2
c $\dfrac{1}{12}$
d 4

5
a Individual response
b Nm C^{-1}

Practice 4

1
a $\dfrac{81}{25}$
b -3
c 4
d $\dfrac{9}{16}$
e $\dfrac{10}{9}$

f $\dfrac{3}{7}$
g 2
h $\dfrac{25}{9}$
i -4
j $\dfrac{1}{9}$

2
a 1
b 9^{21}
c a^5
d a^5

e $\dfrac{y^2}{x^{14}z^{20}}$
f $72t^5$
g $\dfrac{x^{\frac{1}{4}}}{2}$
h $648y^5$

i $9x^4y^5$
j $\dfrac{c^3}{2ab^4}$
k $6a^{\frac{7}{6}}$
l xy^3

m $\dfrac{189x^5y^5z^3}{a}$
n $16g^6h^2$
o $\dfrac{5c^3}{ab^7}$
p $\dfrac{3}{16a^2b^5}$

q $4c^6d^8e^{10}$
r $36cd^{11}$
s $\dfrac{2}{243j^8k^2}$
t $\dfrac{1}{2f^4g^3}$

Practice 5

1
a 2.35×10^4
b 3.658×10^5
c 2.1×10^8
d 3.65×10^6
e 5.69×10^5
f 7.8×10^9

2
a 1 450 000
b 0.002 807
c 9800
d 3 700 000 000
e 0.0 000 506
f 0.00 000 002

3 Smallest to largest: $302\times10^{-6}, 9.83\times10^{-4}, 0.0025, 14, 2.876\times10^2, 7.8\times10^3, 1.42\times10^4$

4 2.99792×10^5 km/s; 9.80665 cm/s^2; 4.5×10^{-5} teslas; 2.25×10^8 m/s; 10^{-9} meters

5
a 10^{100}
b Individual response
c $10^{10^{100}}$
d 7×10^{100}

Practice 6

1
a 6.5×10^6
b 3.43×10^{-1}
c 5.66×10^{-2}
d 7.312×10^5
e 6.916×10^{11}
f 3.724×10^{-4}
g 6.4×10^{-2}
h $6.394\,\overline{230}\,\overline{769}$
i 3.36×10^1
j $4.149\,148\,381\ldots\times10^6$
k 5.776×10^{-5}
l 1.6248×10^7
m $8.89\,097\,744\ldots$
n $2.7\,610\,176\times10^{11}$
o $1.31\,578\,947\ldots\times10^2$
p 9.6824×10^2
q $1.95\,916\,005\,376\times10^{21}$

2
a 1.84×10^{19}
b 3.1536×10^7
c 5.85×10^{11} (to 3 s.f.)

3 a 4.375
   e 5.22×10^4
   b $5.03 \times 10^{-13}\,\text{m}^2$
   c $2.51 \times 10^{-6}\,\text{m}^2$
   d $1.01 \times 10^{-5}\,\text{m}^2$

4 a 22
   b 862 hours or about 36 days.
   c 1.30×10^2 (to 3 s.f.)

5 $4.32 \times 10^4\,\text{km}$

Unit Review

1 a Rational
   b Irrational
   c Rational
   d Irrational
   e Rational
   f Rational
   g Rational
   h Rational

2 a $\dfrac{2}{9}$
   b $\dfrac{1157}{99}$
   c $\dfrac{31}{10}$
   d $\dfrac{22}{9}$

   e $\dfrac{217}{90}$
   f $-\dfrac{431}{500}$
   g $\dfrac{3013}{990}$
   h $\dfrac{6661}{900}$

3 $10^{11}, 10^{10}, 10^9, 10^8, 10^7, 10^6, 10^5, 10^4, 10^3, 10^2, 10^1, 10^0, 10^{-1}, 10^{-2}, 10^{-3}, 10^{-4}, 10^{-5}, 10^{-6}$

4 a $\dfrac{1}{125}$
   b $-\dfrac{1}{27}$
   c -3
   d $\dfrac{9}{16}$

   e $\dfrac{5}{2}$
   f $\dfrac{5}{2}$
   g 4
   h $\dfrac{49}{36}$

5 a $\dfrac{a^7}{b^3}$
   b 1
   c $28m^5n^5$
   d 1
   e $18x^7y$

   f $-\dfrac{2x^9}{y^2}$
   g $-\dfrac{24y}{x^4}$
   h $\dfrac{105}{x^{2c}}$
   i $\dfrac{5m^2}{2n^4}$

6 a 1.2×10^{14}
   b 2×10^3
   c 7.75×10^{-7}
   d $2.\overline{190\,476} \times 10^{-3}$
   e $1.19\,117\,647 \times 10^5$
   f 3.038×10^{-1}

7 a 39.2 days
   b 288 days

8 a 5.8×10^8
   b 5.531×10^{-2}
   c -7.87×10^{-3}
   d 5.627×10^4
   e 4.491×10^{13}
   f 4.973×10^9

9 a 4.842×10^9
   b 2.7158×10^{10}

10 a 8.569×10^{10}
   b Stars greater, by factor 8.17×10^{11}
   c $8.569 \times 10^6\,\text{m}$

11 a $\dfrac{7p^{13}q^5}{r^6}$
   b $\dfrac{v^6}{64u^6}$
   c $\dfrac{1}{a^{10}b^9}$
   d $\dfrac{y^{12}a^7b}{x^5}$

   e $-\dfrac{4a^4c^2}{3b^4}$
   f $\dfrac{6m^{\frac{7}{2}}n^{\frac{5}{2}}}{11}$
   g $-\dfrac{10j^{\frac{19}{6}}}{2g^{\frac{7}{6}}h^{\frac{7}{6}}}$

12 a 10
   b $10^{0.4}$ (≈ 2.51)
   c $10^{2.2}$ (≈ 158)

2 Triangles

Practice 1

1 a $9.2\,\text{mm}$
   b $8.4\,\text{m}$
   c $77.2\,\text{cm}$
   d $24\,\text{cm}$
   e $10\,\text{m}$
   f $8.7\,\text{m}$

2 a Yes, the triangle contains a right angle
   b No, the triangle does not contain a right angle
   c Yes, the triangle contains a right angle

3 a $40\,\text{cm}$
   b $1200\,\text{cm}^2$

4 7.07 cm to 3 s.f.

5 a 5.77 m to 3 s.f.
   b 3.88 m to 3 s.f.

6 90.0 m to 3 s.f.

7 60.2 m to 3 s.f.

8 No: the maximum height of the television can be 13 inches and this corresponds to a television width of $\sqrt{42^2 - 13^2}$ inches = 39.9 inches, which is longer than the width of the cabinet (note that it is also acceptable to provide a similar argument for the comparative heights, or alternatively find the diagonal of the cabinet and show this is less than 42 inches).

9 a 5 **b** 13 **c** 15 **d** 9

10 Work out:

area of semicircle with diameter 3 (*a*) + area of semicircle with diameter 4 (*b*) = $\dfrac{\pi \times 1.5^2}{2} + \dfrac{\pi \times 2^2}{2} = \dfrac{25\pi}{8}$

area of semicircle with diameter 5 (*c*) = $\dfrac{\pi \times 2.5^2}{2} = \dfrac{25\pi}{8}$

so Pythagoras' theorem still applies for semicircles:

the area of the semicircle on the hypotenuse is equal to the sum of the areas of the semicircles on the other two sides

Practice 2

1 a 7.2 **b** 17.8 **c** 9.8 **d** 8.9

2

Original Location	Final location	Actual distance travelled	Distance data was supposed to travel (d)	Did the data end up where it was supposed to? (Y/N)
(2, 6)	(8, 14)	**10**	10	**Y**
(−1, −3)	(2, −1)	**5**	6	**N**
(1, −1)	(−4, 5)	**7.81**	7	**N**
(0, 7)	(−5, −5)	**13**	13	**Y**

3 a i No Hit **ii** Hit **iii** Hit **iv** No Hit

b Individual response (Any pair of coordinates such that the distance between the points is less than 2. e.g. (0,0) and (0,1))

4 11

5 −2 or 14

6 7310 miles to 3s.f.

Practice 3

1 a Not similar

b Similar (corresponding sides are in the same proportion)

c Not similar **d** Not similar

e Not possible to tell; not all right triangles are similar

f Similar (AA postulate)

2 a $\triangle OKL \sim \triangle OMN$ (AA postulate) **b** $\triangle ABE \sim \triangle ACD$, AA postulate; 2.8 cm

$x = 6\,\text{m}$

c $\triangle PQR \sim \triangle STR$, AA postulate; 17.5 cm

d $\triangle ABE \sim \triangle CDE$ **e** $\triangle QPS \sim \triangle SQR$, AA postulate; $x = 15\,\text{cm}$, $y = 25\,\text{cm}$

$x = 0.67\,\text{ft}$

3 a The angle enclosed by the line of sight of the man and the floor, and the angle enclosed by the line from the light to floor and the floor

b Two angles are the same in each triangle, so all three angles are the same. Therefore, the AA postulate can be applied, and the triangles are similar

c Measure the horizontal distance between the man and the surface, and the distance between the surface to find the proportionality of scaling. Then, measure the vertical distance of the man's eye from the floor and scale this proportionally

4 a $\triangle PQR \sim \triangle TRS$, AA postulate

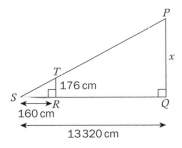

b 14652 cm = 146.5 m (to 1 d.p.)

c When his shadow is the same length as his height, the pyramid's shadow would be the same length as its height

5 Hypotenuse: $\sqrt{200}$ cm
Two other sides: 10 cm

Practice 4

1 a 0.85 **b** 0.21 **c** 0.96 **d** 0.50
  e 0.71 **f** 0.88

2 a 30.0° **b** 48.6° **c** 67.4° **d** 33.1°
  e 39.8° **f** 8.6°

3 a $\sin B = \dfrac{8}{17}$ **b** $\sin B = \dfrac{3}{5}$ **c** $\sin B = \dfrac{12}{13}$ **d** $\sin B = \dfrac{6}{7}$

  $\sin C = \dfrac{15}{17}$ $\sin C = \dfrac{4}{5}$ $\sin C = \dfrac{5}{13}$ $\sin C = \dfrac{\sqrt{3}}{7}$

  e $\sin B = \dfrac{1}{2}$ **f** $\sin B = 0.69$

  $\sin C = \dfrac{\sqrt{3}}{2}$ $\sin C = \dfrac{15.9}{22}$

4 a 29.2 cm **b** 11.5 cm **c** 28.1° **d** 18.6 cm
  e 38.7° **f** 36.9°

5 a 5.80 m to 3 s.f. **b** 4.66 m to 3 s.f.
     Closed length: 2.3 m

Practice 5

1 a 0.469 **b** 0.306 **c** 0.999 **d** 0.675
  e 0.875 **f** 0.695

2 a 33.69 **b** 33.56 **c** 35.10 **d** 70.71
  e 65.80 **f** 73.70

3 a i $\dfrac{4}{5}$ **ii** $\dfrac{3}{5}$ **iii** $\dfrac{4}{5}$ **iv** 1

  v $\dfrac{3}{5}$ **vi** $\dfrac{3}{4}$

  b The side adjacent to angle B is the side opposite C, and the side adjacent to B is adjacent to C

  $\therefore \cos B = \dfrac{\text{Adjacent B}}{\text{Hypotenuse}} = \dfrac{\text{Opposite C}}{\text{Hypotenuse}} = \sin C$ and similar for the second equality

4 a 2.12 m **b** 2.12 m

5 a 18.5 m to 3 s.f. **b** Yes, because the tree is over 18 metres tall

6 34.4° to 3 s.f.

7 Horizontal component: 19.2 ms^{-1}
Vertical component: 16.1 ms^{-1}

Unit Review

1 Individual response

2 a No, the triangles are not similar
  b Yes, the triangles are similar (corresponding angles are equal)
  c Yes, the triangles are similar (corresponding angles are equal)
  d Yes, the triangles are similar (corresponding sides are proportional)
  e No, the triangles are not similar

3 a 1.19 **b** 0.62 **c** 0.26 **d** 0.14 **e** 0.33 **f** 0.49

4 a 51.3°   **b** 53.1°   **c** 26.4°   **d** 49.0°
  e 62.9°   **f** 44.8°

5 a 17   **b** $6\sqrt{6}$ **c** 13.3 (to 3 s.f.)   **d** 12.3 (to 3s.f.)

6 a Postulate AAA   **b** Postulate AAA   **c** Postulate AAA
       14   6 $x = 15$
                                                       $y = 7.5$
                                                       $z = 30$

7 a 5.03 cm to 3s.f.   **b** 58.0° to 3s.f.   **c** 74.8 cm to 3s.f.   **d** 57.3° to 3s.f.

8 73 m

9 a $\sqrt{73}$ **b** $\sqrt{145}$ **c** $2\sqrt{13}$ **d** 5

10 a 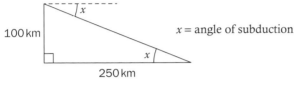   **b** 21.8°

x = angle of subduction

11 a 4 cm   **b** 12 cm²

12 a 12.0 m to 3s.f.   **b** 11.9 m to 3s.f.

13 a 31.3 m to 3s.f.   **b** 82.5° to 3s.f.

14 a

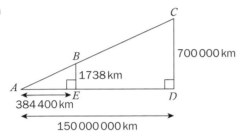

$\triangle ABE \sim \triangle ACD$; AA postulate;

b The angle enclosed by the lines from the top of the Sun to the earth and the Earth the centre of the Sun is 0.53° to the nearest hundredth. The corresponding angle for the Moon is 0.52°. Therefore, the majority of the sunlight will be blocked by the Moon

15 a Individual response (any suitable rectangle with diagonals measuring 80 and 55, such that the area of the larger rectangle is more than double the area of the smaller rectangle)

b

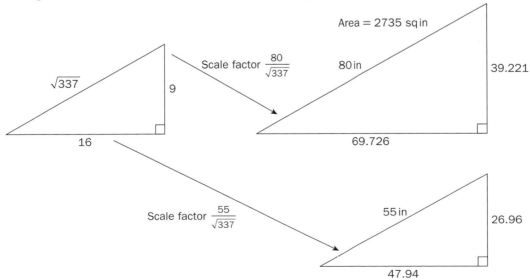

So the 80-inch TV has an area that is more than double the area of the 55-inch TV

16 a Let the pair of points be (x_1, y_1, z_1) and (x_2, y_2, z_2)

Then the distance between the points is

$$\sqrt{(x_1 - x_2)^2 + (y_1 - y_2)^2 + (z_1 - z_2)^2}$$

b i 390.0 light years   **ii** 4.3 light years   **iii** 8.6 light years

c 386.7 light years

3 Linear Relationships

Practice 1

1 a ii   **b** iv   **c** iii   **d** v   **e** i

2 a   **b**

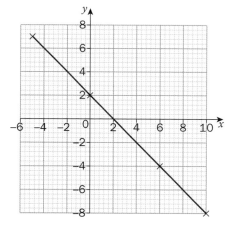

c Individual response

3 a

Year	Cost, $/kWh
2010	1000
2011	875
2012	750
2013	625
2014	500
2015	375
2016	250
2017	125

b, c

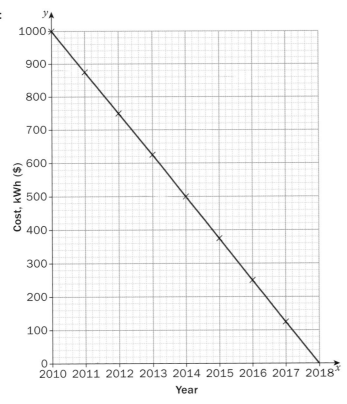

d Individual response
e Individual response

4 Individual response (all parts)

5 **a**

Year	Million tonnes wasted, high-income	Million tonnes wasted, low-income
2018	670	630
2019	1340	1260
2020	2010	1890
2021	2680	2520
2022	3350	3150

b 8710 million tonnes (high income), 8190 million tonnes (low income), from baseline zero in 2017.
c Individual response
d Individual response
e Individual response

Practice 2

1 **a** Linear, 2.5   **b** Linear, −6   **c** Non-linear   **d** Linear, 3

2 **a** $\frac{6}{5}$ **b** $\frac{2}{5}$ **c** −36   **d** $-\frac{10}{21}$ **e** 0

3 No

4 a

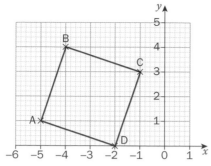

b $(-2,0)$ **c** Individual response

5 a $x = 6;\ y = -5$ **b** $x = \frac{11}{3};\ y = 11$ **c** $x = -3, x = 3;\ y = -\frac{9}{2}$ **d** $x = 14;\ y = 10$

   e $x = -\sqrt{8},\ x = \sqrt{8};\ y = -\sqrt{12},\ y = \sqrt{12}$ **f** $x = 16;\ y = \frac{4}{3}$

6 a

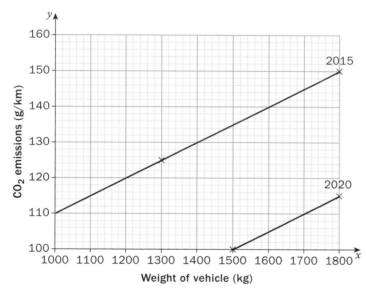

b $0.05(\text{g}/\text{km})/\text{kg}$ **c** $115\,\text{g}/\text{km}$ **d** Individual response

7 a Triangle A: ratio of sides $= \frac{4}{2} = \frac{2}{1} = 2$; triangle B: ratio of sides $= \frac{2}{1} = 2$;

    triangle C: ratio of sides $= \frac{2}{1} = 2$; therefore triangles are similar.

b Gradient of hypotenuse is same for all similar triangles with two sides parallel to axes.

Practice 3

1 a

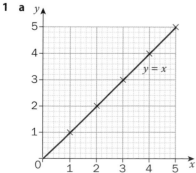

b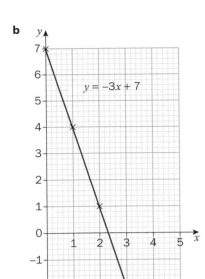

$y = -3x + 7$

c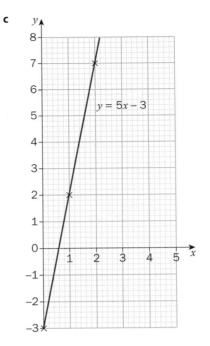

$y = 5x - 3$

d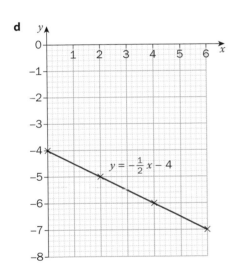

$y = -\frac{1}{2}x - 4$

e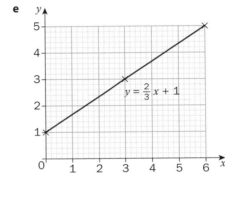

$y = \frac{2}{3}x + 1$

2 **a, f** perpendicular; **b, g** parallel; **b, d** perpendicular; **d, g** perpendicular; **c, h** perpendicular

3 **a**

Time in hours	0	1	2	3	4
Distance in km	0	80	160	240	320

b $80\,\text{km}/\text{h}; x = 0$
c d $280\,\text{km}$
e $y = 80x$; Distance $= 80 \times 3.5 = 280\,\text{km}$
f The train starts from rest and travels $80\,\text{km}$ every hour at a constant speed.

4 **a** Individual response: table or graph. This should start from $(0, 0)$ and go to $(100, 82)$
  b $0.82; x = 0$ **c** $y = 0.82x$ **d** $250\,000$ tonnes   **e** 3.28 billion tonnes

5 **a** $y = 0.6x$ **b** 2.4 billion tonnes   **c** Individual response   **d** Individual response

6 **a** $y = 0.5x$ The cost of one bottle of water.
  b $\$182.50$ **c** 80 **d** Individual response

7 **a** Coal: $y = 80\,000\,000 + 0.2x$ Hydroelectricity: $y = 250\,000\,000 + 0.01x$
  b Coal: cost $= 80\,000\,000 \times 0.2 \times 5000 = \$80\,000\,000\,000$;
      Hydroelectricity: cost $= 250\,000\,000 \times 0.01 \times 5000 = \$250\,000\,050$

c Coal: cost = 80 000 000 × 0.2 × 10 000 = \$160 000 000 000;
Hydroelectricity: cost = 250 000 000 × 0.01 × 10 000 = \$25 000 000 000

d Individual response: slope increases for coal, not for hydroelectricity; y-intercepts unchanged

e Individual response

Practice 4

1 **a** $y = -2x + 4$ **b** $y = \frac{1}{2}x - 2$ **c** $y = \frac{2}{3}x + 3$ **d** $y = 6x + 10$
    e $y = 3x + 9$

2 **a** 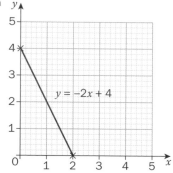   **b**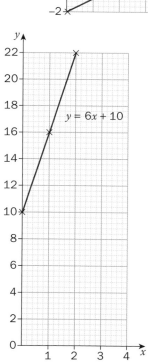

c 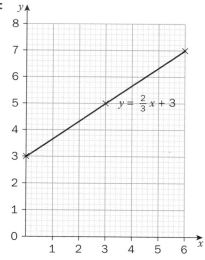   **d**   **e**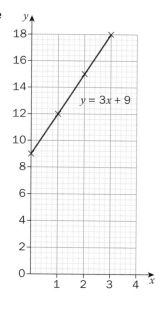

3 **a** $x = 2,\ y = 4$ **b** $x = 4,\ y = -2$ **c** $x = -\frac{9}{2},\ y = 3$ **d** $x = -\frac{5}{3},\ y = 10$
    e $x = -3,\ y = 9$

4 **a, e, g** parallel (gradient $= \frac{1}{7}$); **h, i, k** parallel (gradient $= -2$); **j** perpendicular to **a, e, g** (gradients are -7 and $\frac{1}{7}$;
    b perpendicular to **c** (gradients are 2 and $-\frac{1}{2}$); **d** perpendicular to **l** (gradients are $-\frac{3}{2}$ and $\frac{2}{3}$).

5 **a**

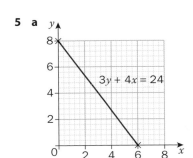

$3y + 4x = 24$

b

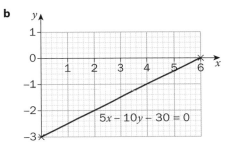

$5x - 10y - 30 = 0$

c

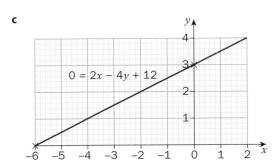

$0 = 2x - 4y + 12$

d

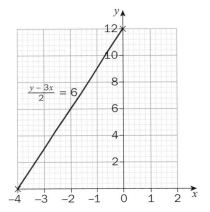

$\frac{y - 3x}{2} = 6$

e

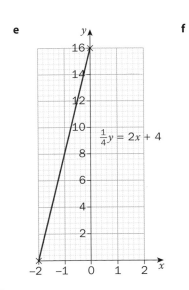

$\frac{1}{4}y = 2x + 4$

f

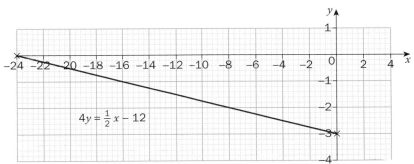

$4y = \frac{1}{2}x - 12$

6 **a**

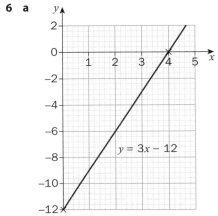

$y = 3x - 12$

b

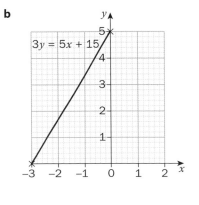

$3y = 5x + 15$

c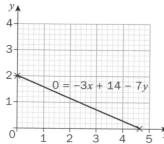

$0 = -3x + 14 - 7y$

d

$y = -x + 4$

e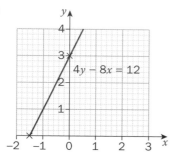

$4y - 8x = 12$

f

$\dfrac{y-2}{3} = -2x$

g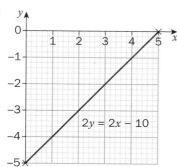

$2y = 2x - 10$

h

$-2 = 3x - \dfrac{1}{2}y$

7 **a** $y = 4x$, $y - 0 = 4(x - 0)$, $4x - y = 0$
Individual response for table and graph

   b Rate of painting; area painted at start.

   c $y = -0.5x + 28$, $y - 28 = -0.5(x - 0)$, $0.5x + y = 28$
Individual response for table and graph; initial time $x = 0$ is taken at 6 pm

   d Rate of cooling, initial temperature

Practice 5

1 **a** $y = 5x + 5$, $y - 5 = 5(x - 0)$, $5x - y = -5$

   b $y = -3x + 7$, $y - 7 = -3(x - 0)$, $3x + y = 7$

   c $y = -2x - 5$, $y + 5 = -2(x - 0)$, $2x + y = -5$

   d $y = \dfrac{1}{2}x + 3$, $y - 3 = \dfrac{1}{2}(x - 0)$, $x - 2y = -6$

2 **a** $y = -4x + 5$ **b** $y = \dfrac{1}{3}x + \dfrac{7}{3}$ **c** $y = -2x + 5$ **d** $y = \dfrac{1}{2}x + 1$

   e $y = \dfrac{4}{3}x + 4$ **f** $y = -2x + 14$ **g** $y = \dfrac{7}{2}x$ **h** $y = \dfrac{7}{3}x + \dfrac{10}{3}$

3 $y = -5x + 6$; $x = 1$

4 **a** $\dfrac{80\,000}{7}$

   b Individual response according to year; following the Tip, the equation is $y = -\dfrac{80\,000}{7}x + 550\,000$ where x is number of years since 2006 and y is number of elephants

   c 2054

5 a $y = 3.2x$ where x is number of years since 1993 and y is sea level rise in mm since 1993
  b 2287

6 a 353 ppm; 399 ppm   **b** 1.7 ppm/year   **c** $y = 1.7x + 336$ **d** Individual response
  e 2135   **f** Individual response

7 a Individual response   **b** Individual response   **c** 17 ppm/ year   **d** −2 ppm/year
  e 2271   **f** Individual response

Practice 6

1 $x = -4$, $x = -\dfrac{3}{2}$, $x = 2$, $y = -3$, $y = \dfrac{5}{2}$, $y = 5$

2 $y = 1$

3 $y = 1$

4 $x = 5$

5 $y = -2$

6 $x = 4$

7 $x = -7$

Unit review

1 a $6, -12$ **b** $-\dfrac{2}{7}, 23$

2 a $x = 4$, $y = 8$ **b** $x = 4$, $y = -5$ **c** $x = 6$, $y = 14$

3 a

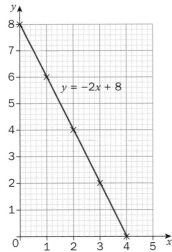

b

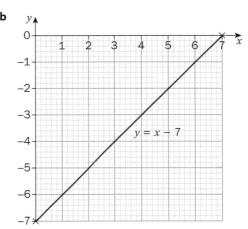

4

Line	Slope	y-intercept	x-intercept	Equation
A	$\dfrac{5}{2}$	−5	2	$y = \dfrac{5}{2}x - 5$
B	0	3	None	$y = 3$
C	$-\dfrac{1}{5}$	2	10	$y = -\dfrac{1}{5}x + 2$
D	Undefined	None	−4	$x = -4$

5 a $y = \dfrac{1}{6}x - 4$ **b** $y = \dfrac{1}{2}x + 2$ **c** $y = \dfrac{1}{8}x$ **d** $y = \dfrac{1}{6}x + \dfrac{2}{3}$
  e $y = 12x - 9$

6 a Undefined   **b** 1   **c** 2   **d** 1

7 a Constant decrease   **b** $L = -0.5d + 100$ **c** 200

8 a

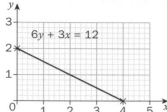

b

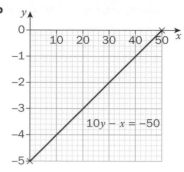

c

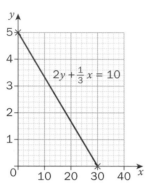

d

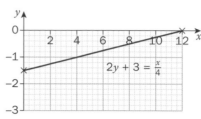

e

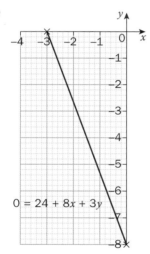

f

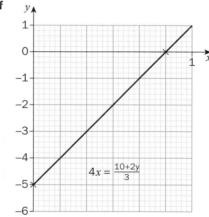

9 a ii   **b** iii   **c** i

10 1

11 $y = -3x + 1$

12 $y = -\dfrac{2}{3}x + 7$

13 $y = \dfrac{1}{3}x - \dfrac{17}{3}$

14 $x = 6$

15 $x = -1$

16 a $y = -0.3x + 1.8$ **b** 2022   **c** Individual response: trend not likely to be linear

17 a $\dfrac{6900}{41} \approx 168$

  b Individual response according to year: equation is $y = -\dfrac{6900}{41}x + 14000$ where x is years since 1975 and y is number of cheetahs

  c 2058

18 $y = \frac{1}{2}x - 3$

19 a 9   **b** −4

20 Yes. Gradient AB $= \frac{2}{3}$; gradient BC $= -\frac{3}{2}$ so perpendicular because $\frac{2}{3} \times -\frac{3}{2} = -1$

21 a −0.321% / year
    b Individual response according to year
    c $y = -0.321x + 18.6$ where x is years since 1991 and y is number of people who are undernourished.
    d 2049
    e Individual response
22 a $y = 0.08x + 19\,000$; $y = 0.03x + 21\,600$
    b 21400; $22500
    c $23\,800$; $23\,400
    d Individual response
    e 52 000 miles If the number miles is less than 52 000 it is cheaper to drive the Ford Focus, but if the number of miles driven is more than 52 000 it is cheaper to drive a zero-emission electric Focus.
    f Individual response

23 a Individual response: values on table should satisfy equation $y = 3.85x$ where x is animal meat in kg and y is feed in kg.
    c Individual response   **d** Individual response   **e** Individual response
    f Individual response   **g** Individual response

4 3D Shapes

Practice 1

1 a 150.8 cm^2 **b** 628.3 cm^2 **c** 703.7 cm^2 **d** 1319.5 cm^2
   e 432.3 cm^2 **f** 2535.2 cm^2

2

Radius (cm)	Height (cm)	Surface area (cm²)
12	18	**2261.9**
10	11	**1319.5**
7	6.9	611.4
8	11.6	985
3	**2.3**	100

3 20 cm straw with a radius of 0.65 cm

4 a $101\pi = 317.3$ m (1 d.p.)   **b** $303\pi = 951.9$ m^2 (1 d.p.)   **c** $606\pi = 1903.8$ m^2 (1 d.p.)

5 a $672\pi = 2111.2$ m^2 (1 d.p.)   **b** 160.5 m^2 (1 d.p.)   **c** 37.3 m^2 (1 d.p.)

Practice 2

1 a Surface Area: $512\pi = 1608.5$ cm^2 **b** Surface Area: $200\pi = 628.3$ cm^2
     Volume: $1536\pi = 4825.5$ cm^3 Volume: $375\pi = 1178.1$ cm^3
   c Surface Area: $36\pi = 113.1$ cm^2 **d** Surface Area: $112\pi = 351.9$ cm^2
     Volume: $28\pi = 88.0$ cm^3 Volume: $160\pi = 502.7$ cm^3
   e Surface Area: 693.2 cm^2 **f** Surface Area: 5170.8 cm^2
     Volume: 1102.7   Volume: 30 283.2 cm^3
   g Surface Area: 7803.7 cm^2
     Volume: 23 411.1 cm^3

2

Radius of base	Height of cylinder	Volume
4.5 cm	0.30 cm	19.4 cm³
3.42 cm	15 cm	550 cm³

3 a $3\,031\,558.4\,m^3$ to the nearest tenth **b** 95.2% to the nearest tenth
  c 808 to the nearest whole number

4 a $34\,494.3\,m^3$ **b** The volume would increase by a factor of 4

5 a $54.6\,m^3$ to the nearest tenth **b** $341\,495\,541$ years **c** $20.1\,m^2$ to the nearest tenth

Practice 3

1 a $301.6\,cm^3$ **b** $102.6\,cm^3$ **c** $2\,120\,575.0\,cm^3$ **d** $67\,858.4\,cm^3$
  e $4.6\,m^3$ **f** $1781.3\,cm^3$

2 a $78\,539.8\,mm^3$ to the nearest tenth **b** $35.4\,mm$ to the nearest tenth

3 a $4084.1\,cm^3$ to the nearest tenth **b** $2884.0\,cm^3$ to the nearest tenth
  c $2111.2\,cm^3$ to the nearest tenth

4 a $609.1\,cm^3$ to the nearest tenth

  b Individual response (e.g. it would use less material than a cylinder of the same height so be cheaper to produce)

5 a $4.7\,cm$ to the nearest tenth **b** $7.9\,cm$ to the nearest tenth

Practice 4

1 a Surface area: $66.0\,cm^2$ **b** Surface area: $452.4\,cm^2$
  Volume: $24.9\,cm^3$ Volume: $639.8\,cm^3$
  c Surface area: $201.2\,cm^2$ **d** Surface area: $75.4\,cm^2$
  Volume: $157.1\,cm^3$ Volume: $37.7\,cm^3$
  e Surface area: $1255.8\,cm^2$ **f** Surface area: $1002.2\,cm^2$
  Volume: $3893.5\,cm^3$ Volume: $1710.2\,cm^3$
  g Surface area: $16336.3\,cm^2$ **h** Surface area: $705.3\,m^2$
  Volume: $150796.4\,cm^3$ Volume: $1508.6\,m^3$
  i Surface area: $54.2\,m^2$
  Volume: $29.1\,m^3$

2 $6283.2\,cm^2$ to the nearest tenth

3 a $1264.9\,cm^2$ **b** $994.4\,cm^2$ **c** $816.8\,cm^2$

4 $1107.6\,m^2$ to the nearest tenth

5 $1230.5\,cm^2$ to the nearest tenth

Practice 5

1 a $160\,cm^3$ **b** $168\,cm^3$ **c** $80\,cm^3$ **d** $1332.9\,mm^3$
  e $1633.3\,mm^3$ **f** $292.5\,cm^3$ **g** $22728.0\,mm^3$ **h** $363.6\,cm^3$

2

Height (cm)	Base side length (cm)	Apothem (cm)	Volume (cm³)
4	6	**5**	**48**
6	16	10	**512**
12	**10**	13	**400**
7.5	20	**12.5**	1000
8	**10.9**	**9.7**	314
3.25	**25.5**	**13.1**	702.15

3 a $1\,384\,190\,m^3$

b Individual response (For example, it wouldn't be possible to use the space right up to the apex or in the corners.)

4 a $74\,m$ **b** Increases by a scale factor of 4

Practice 6

1 **a** Surface Area: 800 cm²
         Volume: 1280 cm³
    c Surface Area: 1065.5 cm²
         Volume: 2048 cm³
    e Surface Area: 84.5 cm²
         Volume: 42.4 cm³
    b Surface Area: 223.3 cm²
         Volume: 167.4 cm³
    d Surface Area: 70.9 cm²
         Volume: 25.0 cm³
    f Surface Area: 864 cm³
         Volume: 960 cm²

2 **a** 374.4 mm²   **b** Yes Each face has the same area so there is an equal chance of landing on each face

3 1.95 m to 3s.f.

4 **a** Surface Area: 4920.8 cm² to the nearest tenth   **b** 58.8% to the nearest tenth

Practice 7

1 **a** 1017.9 cm²   **b** 314.2 cm²   **c** 2463.0 cm²   **d** 235.6 cm²   **e** 6939.8 mm²   **f** 942.5 m²

2 22.0 cm to the nearest tenth

3 22 167.08 cm²

4 5 heating balls each with a radius of 12 cm 10 is 1440π 5 is 2880π

5 **a** The surface area increases by a factor of 4
    b The surface area increases by a factor or 9
    c If the radius of a sphere is increased by a factor of k then the surface area of the sphere increases by a factor of k^2

Practice 8

1 **a** Surface area: 2827.4 mm²
         Volume: 14 137.2 mm³
    c Surface area: 56 410.4 mm²
         Volume: 1 259 833.1 mm³
    e Surface area: 5026.5 mm²
         Volume: 33 510.3 mm³
    b Surface area: 3217.0 cm²
         Volume: 17 157.3 cm³
    d Surface area: 15 843.1 mm²
         Volume: 144 347.8 mm³

2

Radius (cm)	Diameter (cm)	Surface Area (cm²)	Volume (cm³)
15	30	**2827.4**	**14 137.2**
25	50	**7854.0**	**65 449.8**
8.9	**17.8**	1000	**2973.5**
6.2	**12.4**	**483.6**	1000

3 **a** 33.5 m³ to the nearest tenth
    b 1.56 m to the nearest hundredth
    c Individual response e.g. it could sink or burst

4 **a** 50 265.5 m³
    b Individual response e.g. black surfaces absorb heat radiation better than shiny/light colored ones

5 **a** 42.4 cm³   **b** 2077.6 cm³   **c** 5294.4 cm³

6 47.6% to the nearest tenth

7 3

Unit Review

1 **a** Surface area: 294.1 mm²
         Volume: 362.3 mm³
    d Surface area: 1440 cm²
         Volume: 3200 cm³
    b Surface area: 394.1 cm²
         Volume: 735.6 cm³
    e Surface area: 280 cm²
         Volume: 280 cm³
    c Surface area: 1961.6 cm²
         Volume: 5183.6 cm³

2 **a** 33 510.3 m³   **b** 5026.5 m²

3 **a** 61 575.2 cm³   **b** 9324.2 cm²

4 10.0 cm

5 Base: 3 cm by 3 cm; height: 3 cm

6 7.07 inches³ to the nearest hundredth

7 79.6 m³

8 a Surface Area: 97.1 m²   **b** Surface Area: 1306.9 m²   **c** Surface Area: 176.3 m²
    Volume: 62.5 m³   Volume: 4155.3 m³   Volume: 107.3 m³
   d Surface Area: 1309.3   **e** Surface Area: 1680.5 cm²
    Volume: 3152.7 cm³   Volume: 4593.2 cm³

9 Enough to cover 4957.1 m²

10 a 942.62 km²   **b** Approximately 2   **c** Approximately 2

11 6.40 cm to the nearest hundredth

12 13.5 cm

13 a 12.2 m³ to the nearest tenth   **b** 33.2 m²

14 Volume: 183.3 mm³
    Area: 172.8 mm³

5 Bivariate data

Practice 1

1 a Form: Linear   **b** Form: Non-linear   **c** Form: Linear
    Strength: Moderate   Strength: Strong   Strength: Strong
    Direction: Positive   Direction: Positive   Direction: Positive
   d Form: Linear   **e** Form: No pattern   **f** Form: Non-linear
    Strength: Moderate   Direction: No correlation   Strength: Strong
    Direction: Negative   Direction: Negative

2 a Discrete: because only certain values are possible, i.e.the number of tickets and the total cost are integers
   b Total cost
    Justification: the total cost is dependent on the number of tickets
   c Number of tickets
    Justification: this is the quantity that is being varied to observe a change in the total cost
   d

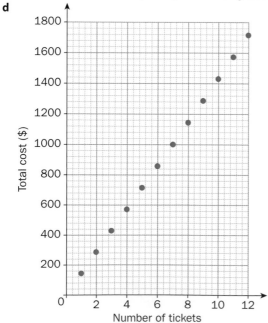

e The data points do line up perfectly; there is a linear relationship so it would make sense to join the points to form a line. (Individual response)

f Form: Linear
Strength: Strong
Direction: Positive

g Denote the number of tickets by n
Denote the total cost by t
Then, $t = 143n$

3 a

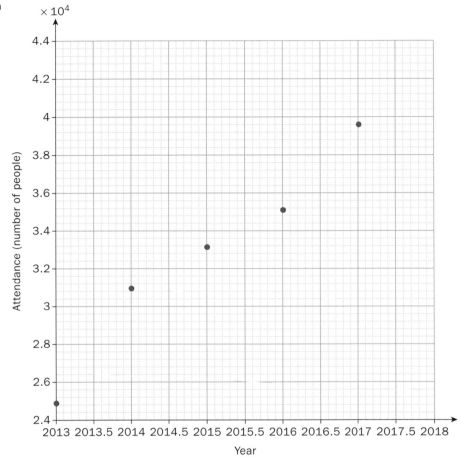

b Form: Linear
Strength: Moderate
Direction: Positive

c Individual response (e.g. between 40 000 and 44 000)

d Individual response (e.g. greater popularity through publicity or word of mouth)

4 a

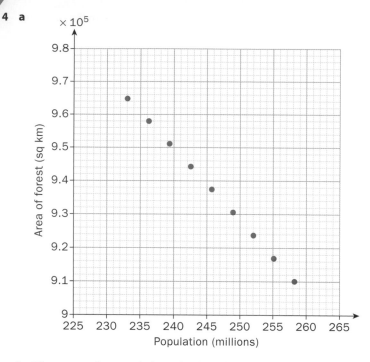

b The greater the population of Indonesia, the smaller the area of remaining forest in Indonesia

c Individual response (student answers should mention that the area of forest is likely to have decreased further)

d Form: Linear
Strength: Strong
Direction: Negative

e Individual response e.g. the scatter plot suggests that there is negative correlation between the area of forest and the population, but this does not mean that it is the increase in population that is causing the decrease in area

5 a

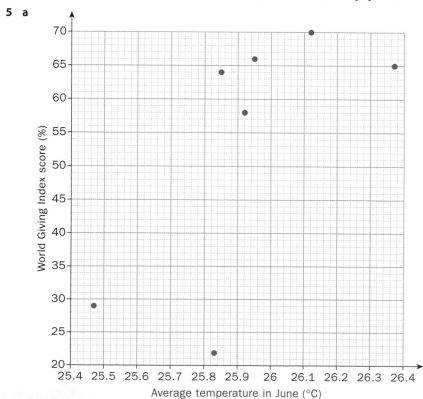

b Form: linear
Strength: weak
Direction: positive

c Individual response The plot suggests that the average temperature has a weak affect on the charitable nature of the citizens of Myanmar

d Individual response e.g. the number of people who are Buddhists as this religion encourages giving

e No; correlation does not imply causation

6 a Generation is categorical; Averages number of job changes is discrete - although decimal values are given, you can only change jobs an whole number of times

b No

c The more recent the generation, the greater the number of times a person changes job by the age of 30 on average

d Individual response   **e** Individual response

f Form: Non-linear
Strength: depends on values given to independent variable
Direction: Positive

Practice 2

1 a $y = -3x$

b $y = \frac{1}{8}x + \frac{13}{8}$

c $y = -\frac{1}{7}x - \frac{29}{7}$

d $y = \frac{11}{6}x - 31$

e $y = 2x + 71$

f $y = -\frac{7}{64}x + \frac{1067}{64}$

2 a $y = -\frac{8}{25}x + 156$ **b** $y = \frac{5}{2}x + 15$

3 a

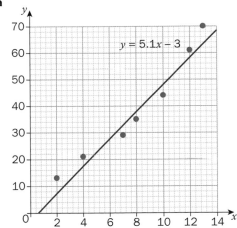

b

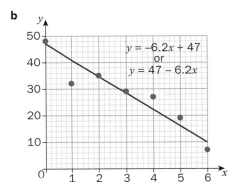

c

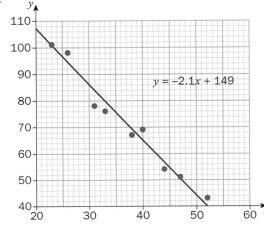

d

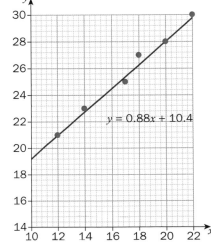

4 Individual response (easiest way to do this is to choose a scatter plot where all data points lie on the same exactly straight line)

5 a

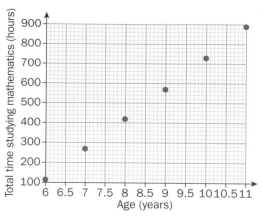
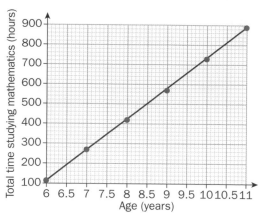

Individual response e.g. $y = 157x - 828$

c $y = 154.429x - 814.476$

d Individual response e.g. The result using technology will have more significant figures

e 1502 to the nearest whole number

f Individual response e.g. There is no data for 15-year-old students so you cannot be sure that the trend will continue

6 a

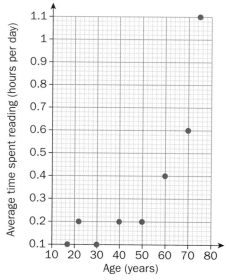
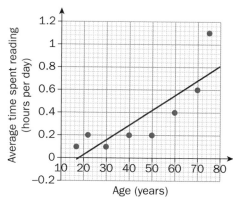

Individual response e.g. $y = 0.01x - 0.2$

c $y = 0.0131x - 0.233$

d Individual response e.g. As the correlation was weak it was not easy to draw the line of best fit

e 75

f Individual response Not very confident: the 70-year-old person in the survey read for 1.1 hours a day

g The older they are, the longer on average they spend reading per day

h Individual response e.g. Older people who are retired have more time in the day to read

i Individual response (e.g. no, they are just more likely to have spare time in which they can read)

7 a

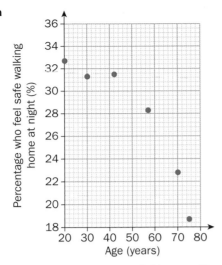

b

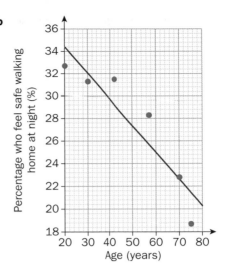

c $y = -0.235x + 39.077$ e.g. $0.24x + y = 39$
d Individual response e.g. The correlation was not very strong so it was not easy to draw the line of best fit
e Individual response depending on age e.g. the percentage for 15-year-olds is predicted to be 35.6%
f Individual response
g The older the person, the less likely they are to be comfortable walking home at night
h Individual response (this is debatable) e.g. Perhaps you could carry out a survey of members of your class or year group and compare the results

Practice 3

1 a No correlation   b Negative, weak   c Positive, strong   d Positive, weak
e Positive, weak   f Negative, moderate   g Negative, moderate   h Positive, moderate
i No correlation   j Negative, strong

2 a 0   b −0.25   c 0.8   d 0.5
e 0.5   f −0.7   g −0.6   h 0.75
i 0   j −0.9

3 Individual Reponses, for example:
a Expected correlation: Positive correlation
Justification: The longer you spend on an assignment, in general, the higher grade you are likely to be awarded
b Expected correlation: Negative correlation
Justification: The greater the number of immunized people, the fewer cases you would expect to see
c Expected correlation: Positive correlation
Justification: The older the person, the more probable it is that at some point in their past they have been involved in a romantic relationship
d No correlation unless more siblings means that the available food needs to be shared between more children, in which case it will show a negative correlation
e Expected correlation: Positive correlation
Justification: the older the adolescent, the greater the expected number of neural connections in the brain
f No correlation

4 Individual responses e.g. a, b strong; c and e moderate; d no correlation or weak; f no correlation

5 a Negative correlation of moderate strength
b Individual response (e.g. between −0.8 and −0.5)
c The greater the number of years in education, the lower the unemployment rate

6 Individual responses, for example
0.8: The age of a person and the total number of books they have read
−0.2: The temperature outside and sales of hot beverages
1: Number of (identical) sweets bought and the total expenditure on that type of sweet
−0.9: Number of hours spent watching TV at home and achievement in a mock exam
−1: Number of (identical) sweets bought and the amount of money you have left available to spend
0.5: The height of a person and the person's weight

Practice 4

1 a 0.543

Positive correlation, moderate

b −0.943

Negative correlation, strong

c 0.976

Positive correlation, strong

d −1.000

Negative correlation, strong (perfect)

2 a The correlation coefficient is −1, implying that the data set is perfectly correlated; the data points all lie on the same straight line

b $y = 51 - x$

c Substituting each value of x gives the corresponding value of y

d No: none of the other correlation coefficients are exactly −1 (or 1)

e a $y = 28.333 - 2.333x$

b $y = 127.286 - 1.571x$

c $y = -7.658 + 4.417x$

f Individual response: Substitute some of the x values from the table into your equation for the line of best fit and check that they give the corresponding y values

3 a Dependent variable: Year

Independent variable: Marriage rate

Justification: we are observing the change in the marriage rate as the year increases

b −0.96

c Exactly the same (if quantity X decreases as Y increases, then quantity Y decreases as X increases)

d −0.96

e For the original data: $y = -0.365x + 747.02$

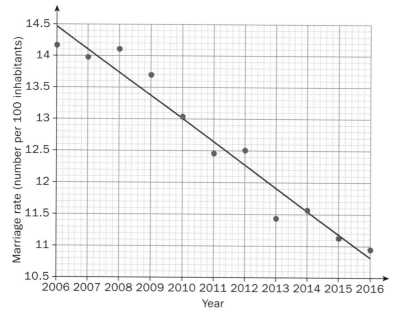

For the switched data: $y = -2.6x + 2044.1$

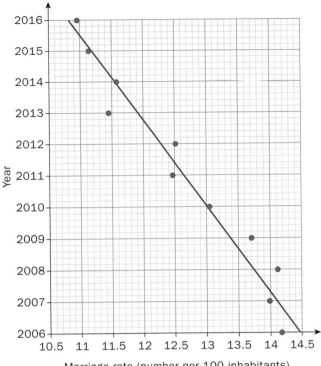

Year (vertical axis)
Marriage rate (number per 100 inhabitants)

f Individual response (e.g. without a good definition of the independent/dependent variable, it is difficult to make meaningful conclusions from data)

4 Individual responses, dependent on the data sets selected

Practice 5

1 None! Correlation does not imply causation.

2 a Individual response (e.g. she is hoping that by moving to a larger house, her reading ability will improve)
   b No (Individual response referring to correlation does not imply causation)
   c Individual response (e.g. wealth: if a family has a larger house then they are likely to be wealthier and therefore provide better learning resources for their children)

3 a The conclusion suggests that the researchers believe there is causation
   b Individual response (e.g. poor work-life balance, or if technology is being used for the purposes of social media, the individuals are subjected to misrepresentations of other lives).
   c Individual response (e.g. time spent socializing in person)

4 Individual response (e.g. if it is raining, then the plants outside are wet, however if the plants are wet, that does not imply it is raining [for example, someone could be watering the plants!])

5 a

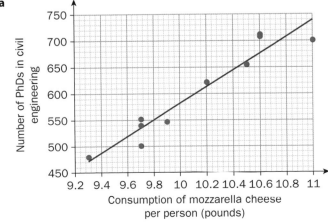

Number of PhDs in civil engineering (vertical axis)
Consumption of mozzarella cheese per person (pounds)

b 0.924

c Form: Linear
Direction: Positive
Strength: Strong

d Equation is $y = -8.25 + 52.5x$; 15.4 pounds per person

e No: correlation does not imply causation and these variables are unlikely to be linked

f Individual response

Unit Review

1 i a Form: Linear
Direction: Positive
Strength: Moderate
b Individual response

iv a Form: Linear
Direction: Negative
Strength: Strong
b Individual response

ii a Form: Linear
Direction: Positive
Strength: Weak
b Individual response

v a Form: Non-linear
Direction: Negative
Strength: Strong
b Individual response

iii a No correlation
b Individual response

2 a

Variables	Independent Variable	Dependent Variable	Sketch of scatter plot
1 Number of times you laugh in a day 2 Age	2	1	Individual response with adequate justification
1 Outside temperature 2 Inside temperature	1	2	Individual response with adequate justification
1 Number of languages learnt 2 Number of nationalities in family tree	2	1	Individual response with adequate justification
1 Son's height 2 Father's height	2	1	Individual response with adequate justification
1 Reaction time 2 Age	2	1	Individual response with adequate justification

b Individual response

3 a −1 and 1 **b** direct, inverse **c** stronger, weaker **d** possible, causation

4 a

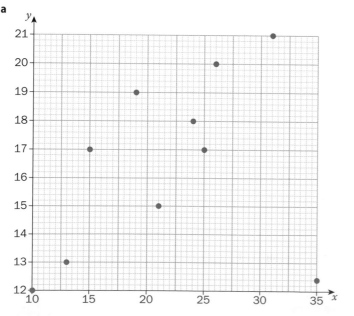

b

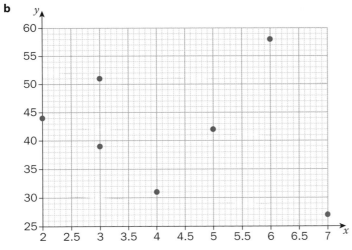

c

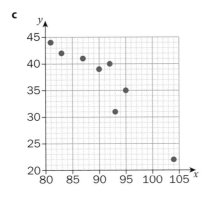

d

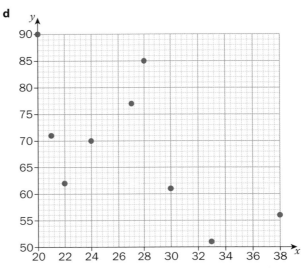

5 a

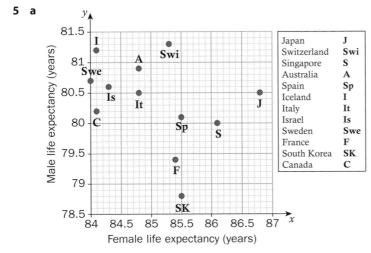

Japan	**J**
Switzerland	**Swi**
Singapore	**S**
Australia	**A**
Spain	**Sp**
Iceland	**I**
Italy	**It**
Israel	**Is**
Sweden	**Swe**
France	**F**
South Korea	**SK**
Canada	**C**

b No correlation
c Comparisons between countries are easier to make using the table

6 a (a)

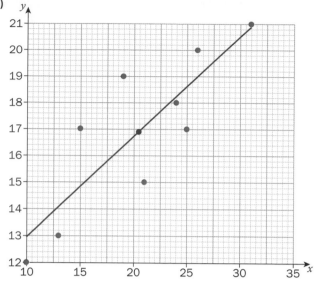

(b)

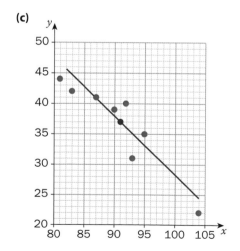

(c)

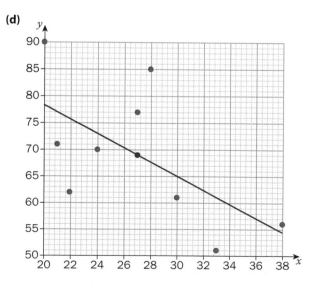

(d)

b Individual response

c (a) $y = 0.377x + 9.175$
    (c) $y = -0.921x + 120.239$

(b) $y = -1.154x + 46.662$
(d) $y = -1.350x + 105.663$

7 Individual response (e.g. The results indicate there is a strong positive correlation between outside temperature and iced tea sales. Causation is likely here; people will prefer to drink cold drinks when it is hot outside. The results also indicate there is a weak negative correlation between outside temperatures and coffee sales. Causation is also likely here; it is likely that fewer people will purchase hot beverages when it is hot outside. Note that the negative correlation for coffee is weaker than the positive correlation for iced tea – this is likely to be due to caffeine addiction tempting people to purchase coffee regardless of the temperature)

8 **a** How many inhabitants there are in a given area of land

  b The sum of the daily temperatures divided by the number of days in a year
Individual response

  c Yes, it has a much higher population density than the other cointries that have a similar average annual temperature

  d No correlation

  e No
Individual response e.g. people don't always choose where they live; it is where they are born

  f Individual response e.g. the cost of living, healthcare, political stability

9 **a**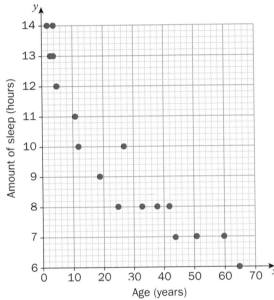

Form: Non-linear
Direction: Negative
Strength: moderate

  b Approximating this relationship using a line of best fit,
$y = -0.117x + 12.758$

  c Slope of the line represents the average change in number of hours of sleep per increase in age of one year
y-intercept represents the hypothetical average number of hours of sleep at age zero

  d Individual response e.g. the equation predicts that a 15-year-old person would have 11 hours sleep

  e 3.4 hours to 1 d.p. no

  f −1.3 hours to 1d.p.
This would not make sense, and hence supports the idea that this is a non-linear relationship

10 **a** Individual response (should be yes)

  b Form: Non-linear
Direction: Positive
Strength: Moderate

  c $y = 230x - 3400$

  d 323; no, this would be too high so the relationship is likely to be non-linear

  e Individual response (e.g. extrapolation likely to be inaccurate)

  f Line of best fit predicts approximately 7400; the actual population is about 6000

  g Individual response e.g. the amount of traffic, the amount of industry, the climate

11 Individual response

6 Geometric transformations

Practice 1

1 a 10 units left, 1 unit down
$(x, y) \mapsto (x - 10, y - 1)$

b 8 units left, 5 units down
$(x, y) \mapsto (x - 8, y - 5)$

2 a i

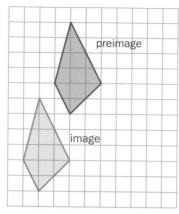

ii

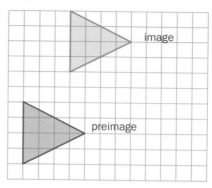

iii

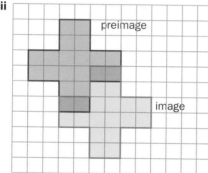

b Yes. The transformation given is equivalent to the simpler transformation 3 units down, 2 units right

c i $(x, y) \mapsto (x - 2, y - 5)$ **ii** $(x, y) \mapsto (x + 3, y + 6)$ **iii** $(x, y) \mapsto (x + 2, y - 3)$

3 Repeatedly apply the transformation
$(x, y) \mapsto (x - 4, y)$
to obtain each subsequent ring (so in total apply the transformation three times)

4 a

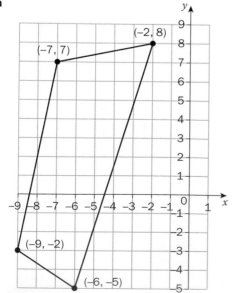

b

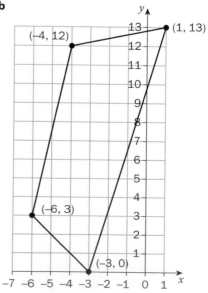

c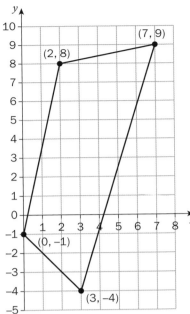

d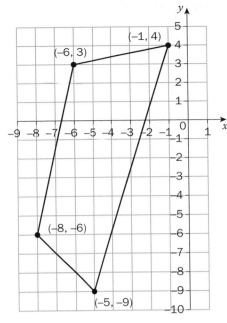

5 a (12,6)   **b** (5,−3)   **c** (7,−11)

6 a (−2,14)   **b** (17,−12)   **c** (4,−1)

7 1L, 3U, 11L, 3D, 3U, 5R, 12U, 5L, 5U, 2D, 25R, 2U, 5D, 5L, 11D, 5R, 3D

8 a Upwards 20 m; sideways by the length of the horizontal diagonal of the diamond
b Yes: the pattern formed by the diamond pieces has no overlaps and no gaps

9 a Horizontally 230 mm; vertically 152 mm; or a combination of 115 mm horizontally and 38 mm vertically
b Yes: the pattern formed by the bricks has no overlaps and no gaps

10 a Individual response e.g. parallelograms and triangles translated horizontally

b There are several tessellation patterns visible

Practice 2

1 a Individual response   **b** Individual response   **c** Individual response

2 a

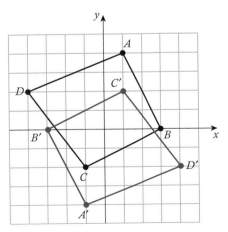

b

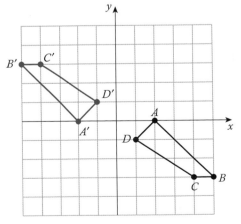

c

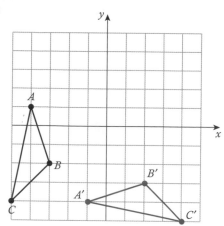

d

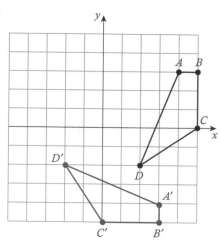

3 a Blue to yellow: $(x, y) \mapsto (x + 2.5, y - 2)$
Blue to black: $(x, y) \mapsto (x + 5, y)$
Blue to green: translation 2 units down, 7.5 units right $(x, y) \mapsto (x + 7.5, y - 2)$
Blue to red: translation 10 units right
  b Individual response, e.g. black to green (x, y) $(x + 2.5, y - 2)$
  c Individual response e.g. from blue to black: rotation 180° center $(-2.5, 0)$

4 a $(-9, 5)$ **b** $(0, -2)$
  c $(8, 0)$
     Rotation of 90° anticlockwise about the origin

5 a $(6, -15)$ **b** $(-13, 11)$
  c $(0, 0)$
     Rotation of 180° anticlockwise (or equivalently clockwise) about the origin

6 a Individual response   **b** Individual response   **c** Individual response
7 a Individual response   **b** Individual response

Practice 3

1 a

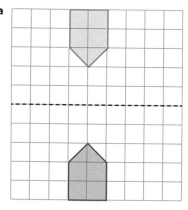

b

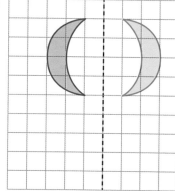

2 a

Either reflect the left/right hand part of the diagram in the vertical dotted line or reflect the top/bottom part of the diagram in the horizontal dotted line

b Rotation of the top/bottom half of the diagram through 180° about the point where the two lines of reflection intersect.

3 a Reflection of left/right hand half of diagram in dotted vertical line

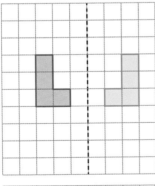

b The fleur-de-lys have been translated horizontally and the symbols in the quarters of the shield have been translated horizontally and vertically

4 a

b

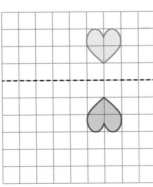

c

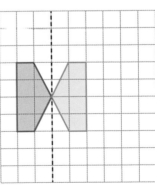

d

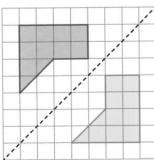

e

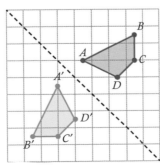

f

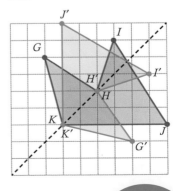

Practice 4

1 a

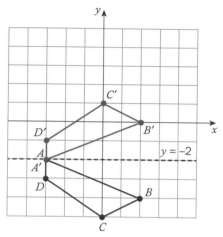

b

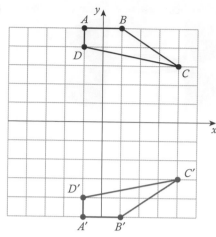

c

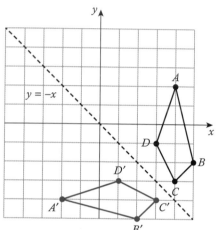

d

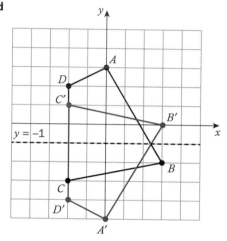

2 a Reflection in the line $y = 2$
b Reflection in the line $y = -x$
c Reflection in the line $y = 0$ (i.e. the x-axis)
d Reflection in the line $x = -3$

3 a $(5, -9)$ **b** $(-2, 0)$ **c** $(0, 8)$

4 a $(6, 15)$ **b** $(-13, -11)$ **c** $(0, 0)$

5 a Individual response   **b** Individual response e.g. the symmetry

6 Hexagons and rhombuses; Individual response

7 a Squares, rectangles and trapezia
b Individual response
c Individual response

8 a $(x, y) \mapsto (-x, y)$ i.e. reflection in y-axis
b $(x, y) \mapsto (-x, y)$ i.e. reflection in y-axis
c $(x, y) \mapsto (-x, y)$ i.e. reflection in y-axis

Practice 5

1 $(12, -28)$

2 $(-4, 9)$

3 $(16, 0)$

4 Scale factor 3

5 a Scale factor $\dfrac{1}{4}$

  b A negative scale factor indicates a dilation and a 180° degree rotation about the centre of dilation

6 Scale factor $-\dfrac{3}{10}$

7 **(4)** $(x, y) \mapsto (3x, 3y)$

  (5) $(x, y) \mapsto \left(\dfrac{1}{4}x, \dfrac{1}{4}y\right)$

  (6) $(x, y) \mapsto \left(-\dfrac{3}{10}x, -\dfrac{3}{10}y\right)$

8

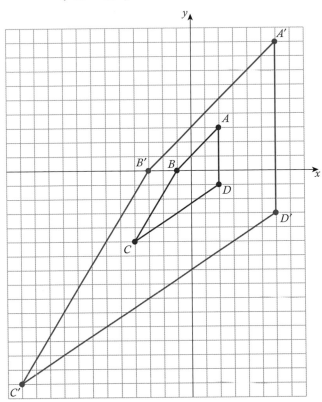

9

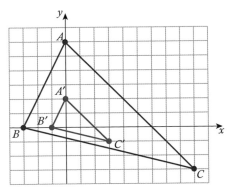

10 a The smallest stars are congruence transformations

  b Stars of different sizes are similarity transformations

11 a Triangles, squares and trapezia are translated, rotated and reflected

  b Yes because there is an overall basic pattern that repeats without overlapping

12 a Repeated dilations of approximately scale factor $\dfrac{1}{4}$

  b No, the pattern cannot be repeated without overlapping

Unit Review

1

Point	Translation up 3 units	Translation left 5 units	Translation down 2 units and right 7 units	Dilation by a scale factor of 4 about the origin
(0, 0)	(0, 3)	(−5, 0)	(7, −2)	(0, 0)
(6, 1)	(6, 4)	(1, 1)	(13, −1)	(24, 4)
(3, −2)	(3, 1)	(−2, −2)	(10, −4)	(12, −8)
(−4, 7)	(−4, 10)	(−9, 7)	(3, 5)	(−16, 28)
(−8, −5)	(−8, −2)	(−13, −5)	(−1, −7)	(−32, −20)

2 a

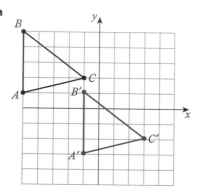

b

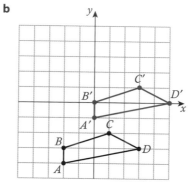

3 No, it does not consist of the same pattern repeated

4 **a** Translation 5 units to the left, 6 units up
   b Rotation of 90° anticlockwise about the origin
   c Reflection in the line $x = 0$ (i.e. the y -axis)
   d Dilation of scale factor 2 centered about the origin
   e Reflection in the line $x = 1$

5 (60, −45)

6 Scale factor $\frac{1}{3}$

7 a

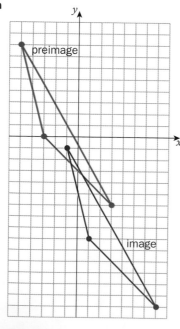

b

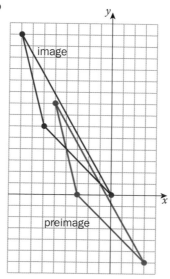

8

Point	Reflection in x-axis	Reflection in y-axis	Reflection in y = x	Reflection in y = −x	Reflection in y = 3	Reflection in x = −4
(0, 0)	(0, 0)	(0, 0)	(0, 0)	(0, 0)	(0, 6)	(−8, 0)
(8, 2)	(8, −2)	(−8, 2)	(2, 8)	(−2, −8)	(8, 4)	(−16, 2)
(7, −6)	(7, 6)	(−7, −6)	(−6, 7)	(6, −7)	(7, 12)	(−15, −6)
(−4, 1)	(−4, −1)	(4, 1)	(1, −4)	(−1, 4)	(−4, 5)	(−4, 1)
(−5, −10)	(−5, 10)	(5, −10)	(−10, −5)	(10, 5)	(−5, 16)	(−3, −10)
(0, −3)	(0, 3)	(0, −3)	(−3, 0)	(3, 0)	(0, 9)	(−8, −3)
(9, 0)	(9, 0)	(−9, 0)	(0, 9)	(0, −9)	(9, 6)	(−17, 0)

9 a

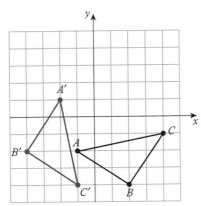

b

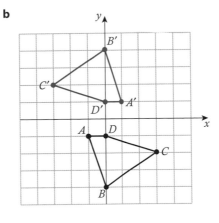

10 a Individual response

b Each square contains reflections and rotations

c Each square has been translated a distance equal to its diagonal horizontally and vertically

11

Point	Rotation of 90 degrees	Rotation of 180 degrees	Rotation of 270 degrees	Rotation of 360 degrees	Rotation of −90 degrees
(0, 0)	(0, 0)	(0, 0)	(0, 0)	(0, 0)	(0, 0)
(8, 2)	(2, −8)	(−8, −2)	(−2, 8)	(8, 2)	(−2, 8)
(7, −6)	(−6, −7)	(−7, 6)	(6, 7)	(7, −6)	(6, 7)
(−4, 1)	(1, 4)	(4, −1)	(−1, −4)	(−4, 1)	(−1, −4)
(−5, −10)	(−10, 5)	(5, 10)	(10, −5)	(−5, −10)	(10, −5)
(0, −3)	(−3, 0)	(0, 3)	(3, 0)	(0, −3)	(3, 0)
(9, 0)	(0, −9)	(−9, 0)	(0, 9)	(9, 0)	(0, 9)

12 a

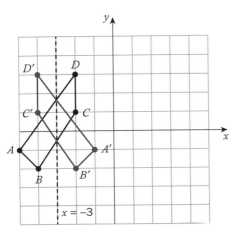

b

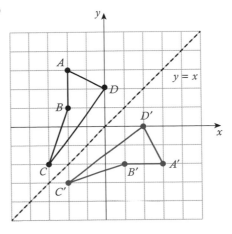

13 a $(x,y) \mapsto (x-5, y-7)$ **b** (x,y) $(0.5x, 0.5y)$ **c** $(x,y) \mapsto (-x,-y)$ **d** $(x,y) \mapsto (x,-y)$

14 a $(14,-7)$ **b** $(-5,12)$
  c $(3,0)$
    Rotation of 90° clockwise about the origin

15 a $(9,0)$ **b** $(4,-14)$
  c $(-5,-3)$
    Rotation of 180° about the origin

16 a Individual response
  b Reflections and rotations
  c Individual response

17 No, the reflection is the transformation $(x,y) \mapsto (-y,-x)$ whereas the rotation is the transformation $(x,y) \mapsto (-x,-y)$

18 a Individual response **b** Individual response **c** Individual response

19 a Rotation of 180° about the origin
  b Reflect in the line $y=0$ i.e. the x-axis

20 a Individual response **b** Individual response

7 Linear systems

Practice 1

1 a $x=-2$ **b** $x=-2$ **c** $x=32$ **d** $x=10$
  e $x=18$ **f** $x=21$ **g** $x=-\frac{1}{2}$ **h** $x=\pm 13$
  i $x=-2$ **j** $x=-1$

2 Individual response

3 Individual response

4 Individual response

Practice 2

1 a $y=-2x+4$ **b** $y=\frac{1}{2}x-2$ **c** $y=\frac{2}{3}x+3$ **d** $y=6x+10$
  e $y=3x+9$ **f** $y=\frac{1}{6}x-4$ **g** $y=\frac{1}{2}x+2$ **h** $y=\frac{1}{8}x$
  i $y=\frac{1}{6}x+\frac{2}{3}$ **j** $y=12x-9$

2 a $x=5, y=\frac{20}{3}$ **b** $x=6, y=-3$ **c** $x=-6, y=3$ **d** $x=-4, y=12$
  e $x=-2, y=16$ **f** $x=24, y=-3$

3 a $x=9, y=12$ **b** $x=1, y=1$ **c** $x=-6, y=-1$ **d** $x=-2, y=-6$
  e $x=-\frac{1}{2}, y=4$ **f** $x=4, y=2$ **g** $x=4, y=3$

4 a

Number of movies	Standard rental, $	Loyalty option 1, $	Loyalty option 2, $
1	6	35	14
2	12	35	18
3	18	35	22
4	24	35	26
5	30	35	30

b

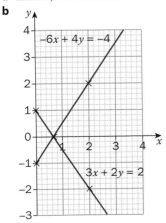

c Option 2 ($10 fee plus $4/movie)
d If you download less than 7 movies a month it is cheaper with Option 2, but if you download 7 or more movies a month it is cheaper with Option 1.
e Individual response

Practice 3

1 **a** One solution   **b** No solutions   **c** No solutions
   d Infinitely many solutions   **e** Infinitely many solutions   **f** One solution
   g No solutions   **h** Infinitely many solutions   **i** One solution
   j No solutions   **k** One solution   **l** No solutions

2 **a** $k = -2$ **b** All $k \neq -2$

3 **a** Individual response: values for the two options must satisfy equations $y = 0.75x + 90$ and $y = 5x$ respectively, where x is number of days and y is cost
   b Individual response based on part **a**
   c For 1 month, if you use the facilities 6 times, membership would cost $94.50 and 'pay as you go' would cost $30. If you used the facilities for 6 times a month for 4 months, membership would cost $108 but 'pay-as-you-go' would cost $120, so it would take 4 months.
   d For 1 month, if you use the facilities 4 times, membership would cost $93 and 'pay as you go' would cost $20. If you used the facilities for 4 times a month for 5 months, membership would cost $102 but 'pay-as-you-go' would cost $100, so you would need to use the facilities for more than 5 months for membership to be the better option.
   e Individual response
   f Membership would be the better option as it will always be cheaper.
   g Make cost per day same with or without joining fee

4 **a** Linear, one solution
   b

$$-6x + 4y = -4$$

$$3x + 2y = 2$$

   c $-6x - 4y = -4$ This is the same equation as $3x + 2y = 2$ if you divide each number by 2 and rearrange.

d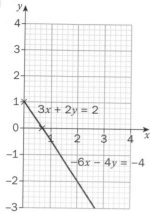

5 Individual response

6 Individual response

Practice 4

1 a $x = 1, y = -3$ **b** $x = 7, y = 1$ **c** $x = \frac{1}{5}, y = 2$ **d** $x = \frac{6}{7}, y = -\frac{6}{7}$

  e $x = 2, y = 2$ **f** $x = -5, y = 1$

2 a $x = 1, y = 1$ **b** $x = -1, y = 4$ **c** $x = \frac{1}{2}, y = \frac{3}{2}$ **d** $x = -2, y = 1$

  e $x = -2, y = -2$ **f** $x = 2, y = \frac{1}{2}$

3 a You earn \$60 for each case you sell so $y = 60x$ where x is the number of cases sold; The initial cost is \$1300 and each case costs \$30 to make so $y = 30x + 1300$

  b 44   **c** \$2640   **d** After selling 44 covers

4 Individual response

Practice 5

1 a $x = 2, y = -1$ **b** $x = \frac{22}{19}, y = \frac{5}{19}$ **c** $x = -2, y = 0$ **d** Infinitely many solutions

  e $x = 4, y = 1$ **f** $x = 2, y = -1$ **g** $x = 4, y = -6$ **h** No solution

  i $x = 4, y = 1$ **j** $x = 2, y = 0$

2 a Individual response

  b Individual response; the ones where elimination suggested that there were no solutions or where there were infinitely many solutions could be confirmed by graphing because there would not be any point where the two lines intersect.

3 a Individual response

  b Individual response; the ones where the coefficient of x or y is 1 are easier to solve by substitution.

Practice 6

1 18

2 13

3 \$600 at 8%, \$400 at 9%

4 3 classrooms, 17 girls

5 45

6 10 sheep, 14 beehives

7 \$6000 at 9%, \$2000 at 10%

8 Adults £5, children £2

9 \$3000

10 Adults $40, teenagers $8

11 a 6400 **b** Rs 3 040 000

12 $2x - 9y = -41$

13 $(3, 4)$

14 a 2.13 hectares **b** 29 000 THB **c** Individual response
  d 3.50 hectares; they would need to increase the number of hectares of their farms in order to continue making a profit.

Unit review

1 a $x = 2$ **b** $x = -6$ **c** $x = -12$ **d** $x = 5$
  e $y = -42$ **f** $y = 10$

2 a $x = 3, y = 1$ **b** $x = 0, y = -6$

3 a $x = 3, y = -5$ **b** $x = -\dfrac{5}{11}, y = \dfrac{164}{11}$

4 a $x = -\dfrac{1}{2}, y = 4$ **b** $a = 4, b = 2$ **c** No solution **d** $x = 2, y = 3$
  e $x = -3, y = -5$

5 a One solution **b** Infinitely many solutions **c** No solution

6 $x = -5$

7 $-6, -5$

8 Square; all the sides are 30 cm long.

9 $x = 14, y = 8$

10 30 coins; 9 nickels, 18 dimes and 3 quarters

11 a

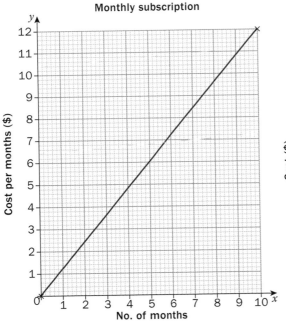

  b 9 **c** Individual response

12 6 teachers, 36 students

13 AU$15000 at 3%, AU$5000 at 8%

14 910 meals that are paid for and 800 free meals so a total of 1710 meals.

15 16 years

16 100 nets, 40 solar kits; 60 medicine kits, 120 blankets.

Index